Shifting Sands

QUARRY BOOKS

AN IMPRINT OF
INDIANA UNIVERSITY PRESS
Bloomington & Indianapolis

Shifting SANDS

The Restoration of the Calumet Area

Kenneth J. Schoon

KENNETH J. SCHOON

This book is a publication of

Quarry Books
an imprint of

Indiana University Press
Office of Scholarly Publishing
Herman B Wells Library 350
1320 East 10th Street
Bloomington, Indiana 47405 USA

iupress.indiana.edu

This book is printed on acid-free paper.

Manufactured in Korea

Cataloging information is available from the Library of Congress.

ISBN 978-0-253-02295-0 (cloth)
ISBN 978-0-253-02340-7 (ebook)

1 2 3 4 5   21 20 19 18 17 16

Contents

Acknowledgments

I AM GRATEFUL FOR THE SUPPORT AND CONTRIBUTIONS OF THE following individuals: Dustin Anderson, Erin Argyilan, Peter Avis, Nicole Barker, Geof Benson, Eric Bird, Scott Bocock, Eric Bottger, Mark Bottger, Lee Botts, whose idea it was to write this book, Mark Bouman, Joel Brammeier, Brian Breidert, Casey Bukro, Brad Bumgardner, Jennifer Caddick, Dorreen Carey, Kelly Carmichael, Young Choi, Candace Clark, Peter Clevering, Bradley Cook, Carole Cornelison, Spencer Cortwright, Steve Coxhead, Erin Crofton, Bob Daum, Therese Davis, Rick DeChantal, Tom Desch, Tom Easterly, Alicia Ebaugh, Eric Ehn, Jim Erdelac, Jeff Farkas, John Fekete, Meghan Forseth, Meredith Gramelspacher, Dale Heinze, Walt Helminski, Gregg Hertzlieb, Tricia Hodge, John Hodson, Eva Hopkins, Amber Horbovetz, Paula Isolampi, Elizabeth Johnson, Brian Kallies, Matthew Keene, Tom Keilman, Anne Koehler, Gayle Kosalko, Anicia Kosky, Kris Krouse, Barb Labus, Paul Labus, Carolyn Lohman, Mark Loomis, Kathy Luther, Richard Lytle, Diana Mally, Jeff Manuczak, Steve McShane, John Mengel, Nick Meyer, Susan MiHalo, Marie Min, Peg Mohar, Kelly Mullaney, Paul Nelson, Caitie Nigrelli, Kelly Nissan-Budge, Ron Novak, Michele Oertel, Noel Pavlovic, Trent Pendley, Aaron Pigors, Dan Plath, Jolice Pojeta, Heather Pritchard, Herb and Charlotte Read, Mark Reshkin, Jeanne Robbins, Ronald Robbins, Jeanette Romano, Mike Ryan, Scott Sandberg, Carrie Sanidas, Rana Segal, JoAnne Shafer, Steve Shook, Candice Smith, Ashley Snyder, Russell Snyder, Donna Stuckert, Jim Sweeney, Kim Swift, Ellen Szarleta, Damon Theis, Gayle Tonkovich, Kim Torp, Ron Trigg, Clay Turner, Marcy Twete, John Watkins, Cindy Watson, Sarah Weaver,

Sandi Weindling, David Wellman, Kristin Wiley, Patricia Wisniewski, Don Woodard, and special thanks to Peg Schoon for copyediting the work, compiling the index, and putting up with my fixation on this effort.

I am appreciative of all the persons and institutions that made their photographs available for inclusion in this book. Donors are listed in the captions. Special appreciation goes to Ron Trigg, who took most of the photographs attributed to Shirley Heinze Land Trust. Illustrations whose captions list no donor were taken by me.

Finally, I am grateful to the Discovery Alliance composed of the Legacy Foundation, the Porter County Community Foundation, the Unity Foundation of LaPorte County, and the Crown Point Community Foundation. Its financial support allowed for a larger page size and full color illustrations throughout the book.

The material in this book has been carefully researched and reviewed; any potential errors that remain are of interest. Followers of environmental or local history who have information that conflicts with that given here are invited to contact the author.

LEGACY
FOUNDATION

Lake County's Community Foundation

PORTER COUNTY
COMMUNITY
FOUNDATION

Unity Foundation
of La Porte County

CROWN POINT
community
foundation™

Just as it is particularly blessed, Northwest Indiana is particularly challenged to achieve a cleaner, safer, richer environment and a sustainable balance between nature and the built environment.

ONE REGION, *Northwest Indiana Profile*

Part One

The Unrestricted Use of Natural Resources

*H*istorian Powell Moore called the Calumet Area "Indiana's Last Frontier." While it was the first part of Indiana to be explored by Europeans, it was the last to be settled and "tamed."

Water has played a big part in determining its fate. It was through the waters of Lake Michigan that French explorers and voyageurs first came into the area in the late seventeenth century. It was the waters of the Kankakee River and its marshland that prevented American pioneers from moving north into the area in the late eighteenth and early nineteenth centuries. It was again by water that the first settlers, Joseph and Marie Bailly, entered the area, and they established their homestead on the waters of the Little Calumet River.

When Indiana became a state in 1816, most of its citizens lived in its southern portion. John Tipton, an early state official who surveyed the Illinois-Indiana border in 1821, described the dune area as ponds, marshes, and sand hills that "can never admit of settlement nor never will be of much service to our state."[1] As late as 1890, a writer in the *Chesterton Tribune* (which would later extol the beauty of the Dunes) described Duneland as "the most godforsaken place in the State of Indiana."[2]

Slowly, however, American pioneers moved into what would soon be LaPorte, Porter, and Lake Counties and recognized their bounty. Slowly the wild landscape, except for the sand hills near Lake Michigan and the wetlands along its rivers, was turned into farmland. When the railroads arrived in the 1850s speculators and enterprising businessmen started investing heavily in the area.

Sand was mined, hauled away, and sold to municipalities, refractories, glass makers, and the railroads themselves. Clay was mined and turned into bricks. Ice was cut in the winter, hauled to distant cities in spring and summer, and sold to businesses and residents alike. Dunes were leveled. Wetlands were drained. Mills were built near sources of abundant water. Wastes were released into the air, piled on the ground, and dumped into the Calumet Area's slow-moving rivers.

The Calumet Area became an industrial giant. If Chicago was the "City of the Big Shoulders," then the Calumet Area was part of the mechanism that made those shoulders big. When it was built in 1974 the Standard Oil Building (today the Aon Center) was the tallest building in Chicago but its refinery was in Whiting. The landmark stainless steel–clad Inland Steel Building was earlier built in Chicago, but its mills were in East Chicago. Chicago had steel mills before Indiana did, but when they all closed, the Indiana mills remained open.

But all this activity created an extremely polluted region. The Grand Calumet River was named the most polluted in America. The air quality was the worst in the state. It was time to change direction.

Calumet Beginnings and the Birth of American Ecological Science

1

IT CAN BE SAID THAT THE GLACIERS MADE LAKE MICHIGAN, Lake Michigan made the beach, and the wind made the Dunes. Although there is much more along the South Shore of Lake Michigan than just the Dunes, the Dunes are what makes this part of the natural world spectacular and unique. They have inspired artists, hikers, and scientists. They are the jewels of the South Shore. They brought Henry Chandler Cowles to the area to study them and their plant life. And by doing just that, he justified the theory of succession and initiated ecological science in this country.

THE EFFECTS OF THE GLACIERS

Although the glaciers have been gone from Northwest Indiana for thousands of years, much of what they formed when they were here remains and has affected the area ever since. Roughly seventeen thousand years ago, the Lake Michigan lobe of the glacier invaded the Calumet Area and deposited huge amounts of sediment along its edge, forming the ridges and hills known today as the Valparaiso Moraine (vm on the map below). The Valparaiso is the largest and highest of the moraines in the Calumet Area, and together with the smaller Tinley/Lake Border Moraines (tm and lbm on the map below), it forms one of the dominant landscapes of the area. It gets its name from the city of Valparaiso, where the moraine is narrower, higher, and steeper than in places to its west.

Later on, the glacier melted back, readvanced, and deposited the sediments that made the Tinley/Lake Border Moraines on the lakeward

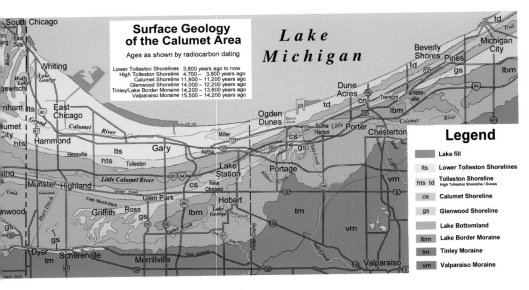

Surface geology of the northern Calumet Area.

flank of the Valparaiso Moraine. Although many moraines were created by glaciers in what are now Indiana and Illinois, the Valparaiso Moraine is significant because it was built upon the top of the Eastern Continental Divide, which separates all rivers and streams that flow north and east to the North Atlantic from those that flow west and south to the Gulf of Mexico. Thus the divide and the moraines form the natural southern end of the Lake Michigan drainage basin.

LAKE MICHIGAN'S ANCIENT SHORELINES

What is now Lake Michigan (but what had earlier been called Lake Chicago) was formed about 14,500 years ago, when the glacial lobe that had entered Northwest Indiana retreated from the Tinley Moraine.[1] Much of the meltwaters from that ice were then trapped between the glacier to the north and the U-shaped Tinley and Valparaiso Moraines.

The Glenwood Shoreline (gs on the map) was the first and is the highest (at 640 feet above sea level) of Lake Michigan's ancient shorelines. During the Glenwood phase a long sand spit, extending from Portage to Griffith, formed parallel to the shoreline. North of this shoreline is the flat former lake bottom containing extensive deposits of clay.

The history of the lake is quite complicated, but the important stages are described here. The Glenwood phase ended about 12,200 years ago when the melting glacier retreated past the Straits of Mackinac. As the straits were then at a significantly lower elevation than the Chicago Outlet, the lake water rapidly drained to the north, the lake level dropped by a large amount, and its waters receded from the shore. The Glenwood phase was over.

The Calumet Shoreline (cs on the map) formed about 11,800 years ago when a glacial advance again blocked the Straits of Mackinac and the lake level rose to 620 feet above sea level. In Lake County the Calumet Shoreline formed several miles north of the Glenwood. In Porter and LaPorte Counties it developed quite close to the older Glenwood Shoreline. In certain places Calumet dunes even bury those of the Glenwood. Lake Michigan's Calumet phase lasted about 600 years, ending roughly 11,200 years ago when once again the glacier retreated past the Straits of Mackinac. The lake waters again drained to the north and the lake level once again dropped. The Calumet phase was over.

The Tolleston Shoreline, the third and youngest of Lake Michigan's shorelines, was formed many years later when the lake level had risen to 605 feet above sea level. It was named the Tolleston Shoreline because it is so prominent in Tolleston, now a part of the city of Gary.

The High Tolleston Shoreline (from Calumet City to Portage, hts on the map) today is a thirteen-mile-long curved ridge, north of and parallel to the Calumet Shoreline. The Lower Tolleston Shorelines (from Chicago to Miller, lts on the map) were composed of more than 150 rather low parallel beach ridges. Long, low swales that often contained standing water lay between the ridges. This landscape is often called "dune and swale" topography, but it might better be called "beach ridge and swale" topography. This rare landscape can today be seen in the several nature preserves in western Gary and Hessville in eastern Hammond. The longest of these narrow dune ridges still remaining can be seen in Hessville's Gibson Woods. The Tolleston dunes (td on the map) in Porter and LaPorte Counties, the result of prevailing northwest winds, are the tallest dunes in Indiana.

Lake Michigan's water level fluctuates with the change of seasons and through drought and periods of above-average rainfall. In a typical year, the lake level rises approximately twelve inches in the spring and summer and then in autumn and winter falls back down to its late winter low level.[2]

Because the Dunes are situated between the hardwood forests of the east and the prairies of the west, Duneland contains plants from both regions. Because wetlands lie between the Dunes, Duneland also contains wetland plants. Because the Dunes are sandy, Duneland has some desert plants from the southwest. Because the glaciers brought seeds down from the north, Duneland has some plants from the north. In fact, the Indiana Dunes National Lakeshore, with more than 1,574 different species of plants, has a greater number of different plant species than any other national park of similar size and more than most of the large parks such as Yellowstone and Yosemite.

THE GRAND AND LITTLE CALUMET RIVERS

The evolution of the three Calumet Rivers was quite convoluted. In presettlement times, today's Grand and Little Calumet Rivers were one long river that began in LaPorte County and flowed west to Blue Island, Illinois, where it crossed the Tolleston Shoreline, then flowed back into Indiana and emptied into Lake Michigan at what is now Marquette Park in Gary. Being for the most part parallel to the Lake Michigan shoreline, it had a low gradient and its waters flowed rather slowly. Close by, but separate in those days, was a "little" Calumet River that connected Lake Calumet to Lake Michigan.

Someone, or a group of people, perhaps over time but definitely before 1805, dredged a low but dry channel between the two Calumet Rivers. This may have been done by Indians or perhaps French fur traders (see General Hull's map below). In any event, during an 1805 flood, river water poured through this low channel from the original Calumet River into that "little" river. The fast-moving river water naturally eroded the sandy channel as it flowed, resulting in a new route for the longer Calumet River. In effect, this action split the Calumet River in half, with the waters flowing through the channel and then down that "little" Calumet

Duneland has plants from the north, east, south, and west. It is often
called the Central Forest-Grasslands Transition Ecoregion.

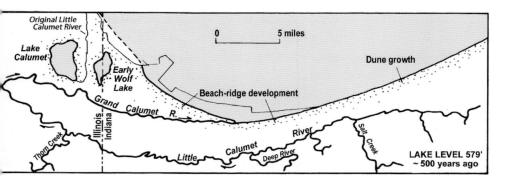

The Calumet Rivers before the 1805 flood connected the original Calumet
River to the "Little" Calumet River west of Wolf Lake. *Chrzastowski and
Thompson, "Late Wisconsinan and Holocene Geologic History," 412.*

General William Hull's 1812 map. *Brennan,* Wonders of the Dunes, *inside back cover.*

River to its mouth, emptying into Lake Michigan at what is now Chicago's Ninety-Fifth Street.[3]

By the mid-1800s the old mouth at Miller Beach had dried up and the nearly flat Grand Calumet changed directions on its own and started flowing west. The original "little" Calumet, now in Chicago, is today called the Calumet River. The lower original river (the north branch) is now the Grand Calumet River, and the upper original river (the south branch) is now the Little Calumet River.

The rivers have had three other diversions since 1805. In 1922 the Little Calumet was linked to the Illinois River system (and thence to the Mississippi River) via the Cal Sag Channel. In 1925, the Grand Calumet was linked directly to Lake Michigan via the Indiana Harbor Ship Canal. Then in 1926 Burns Ditch was dug, connecting the upper Little Calumet River with Lake Michigan. Whereas before 1805 the Calumet River had one mouth, the Calumet Rivers today have four.

All three Calumet Rivers were crystal clear when first seen by area pioneers. They provided those early settlers, and the Potawatomi before them, with wholesome food.

Today the headwaters of the Grand Calumet River are at Gary's Marquette Park Lagoons. From there, the river flows westward to East Chicago where it is diverted and flows north to Lake Michigan through the Indiana Harbor Ship Canal. The section of the river west of the junction has little flow: the short eastern portion flows eastward to the canal, while the lower western portion flows to the Cal Sag Channel in Illinois.[4] The eastern section of the river in Gary is today a manmade channel; its only tributaries are sewer discharge pipes. The Grand Calumet River is much cleaner today than it was thirty years ago, but not as clean yet as it was in 1850.

The Little Calumet River, which begins in the hills of western LaPorte County, extends westward into Illinois. At Riverdale it still makes a hairpin turn and goes north and eastward to its junction with the Grand Calumet River. The Little Calumet has two outlets: one through Burns Ditch in Porter County, and a second where the river empties into the Calumet Sag Channel.

From Thorn Creek in Illinois to Sand Creek in Chesterton, all the streams that flow north from the Valparaiso Moraine flow into the Little

GRAND CALUMET RIVER AT BROADWAY, GARY.   JUNE 1906.

The Grand Calumet River on a postcard in 1906, before the new
channel was dug. Note the crane in the background.

Calumet River. When their waters enter the river their rate of speed
decreases because of the very low gradient of the river. Thus the Little
Calumet has always been prone to flooding. After heavy rains, this usually small-looking river can overflow its banks and flood nearby fields as
well as residential and other properties.

HENRY CHANDLER COWLES AND THE BIRTH
OF AMERICAN ECOLOGICAL SCIENCE

The eastern portion of the Indiana Dunes is a great place to study the
effects of the ages of landscapes. It is in this area that the glacial moraine,
three ancient shorelines, and the current shoreline of Lake Michigan can
all be found within a span of a bit more than one mile. At the south end
is the fourteen-thousand-year-old morainal landscape made by glacial
ice; to the north is an active beach which may have dune ridges formed
within the last two decades.

Dr. Henry Cowles spent much of his career studying plant life in the
Indiana Dunes. He first came as a graduate student in the mid-1890s.
He was then fascinated by the differences in plant populations along the
various ancient shorelines described above and he realized that time, or
landscape age, was an important factor in determining which plants grew

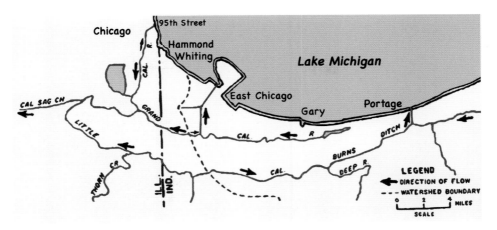

Streamflow directions of the three Calumet Rivers. *Adapted from Simmons et al.,* Information Summary, *fig. 32.*

in which places. As he had a background in both botany and geology, he could recognize relationships between the two disciplines. His work provided solid evidence to support the rather new concept of "succession," or predictable change in plant life over time. Cowles defended his dissertation, entitled "An Ecological Study of the Sand Dune Flora of Northern Indiana," in 1898 and received his PhD in both geology and the rather new field of botany. The next year he published his work. Cowles did not invent the idea of succession, but his work in the Dunes provided a convincing contribution to this very new idea. He was able to show how topography, or the shape of the land, influenced succession, and he was also able to find examples of botanical change over time. Cowles, who stayed at the university for his entire career, then began to take his students to the Dunes.[5]

In 1911, Cowles, who was by this time the chairman of the university's Botany Department, attended a month-long International Phytogeographical Excursion of the British Isles along with several well-known European scientists. This trip was so beneficial to him that upon returning to Chicago he began planning a similar meeting of international scientists in the United States.

The Excursion in America was held in the summer of 1913. Before planning the schedule, he asked those scientists who had committed to attending what parts of the United States they'd like to see. Although the

Henry Chandler Cowles (*left*), Sir Arthur Tansley, and others relax during
a tour of the Dunes. Sir Arthur was the president of the Ecological Society
and founder of the *Journal of Ecology. Dunes Learning Center.*

responses were varied, four sites appeared on practically all the lists: the
Grand Canyon, Yosemite, Yellowstone, and the Indiana Dunes.

The group spent three days in the Dunes area, taking the train from
Chicago and trekking from various stations to see the current and ancient shorelines, the high dunes, the wetlands, and various ecosystems.
One of the areas they visited was the Mineral Springs wetland, which
today is named Cowles Bog.

Marquette and the Marquette Plan

2

Make no little plans; they have no magic to stir people's blood and probably themselves will not be realized. Make big plans; aim high in hope and work. . . . Let your watchword be order and your beacon beauty.

Daniel H. Burnham, *Plan of Chicago*

FATHER JACQUES MARQUETTE AND HIS TWO COMPANIONS were the first known Europeans to travel through the Calumet Area. Before that, it is quite possible that French-Canadian *coureurs de bois* (runners of the woods) came to this area and traded with the Indians. The Calumet Area, whose waters and wetlands housed vast numbers of beaver, muskrat, and mink, would have been a tempting source of income to these men. However, because their trade was by French law illegal, they made no written records of their travels. Things changed in 1681 when the French government in Quebec decided to license this trade. That act initiated the era of legal trade by men who were then called voyageurs.

Father Marquette is practically a folk hero in the Lake Michigan area, in much the same way that Daniel Boone is in the Ohio Valley. Marquette arrived about one hundred years earlier than Boone, and both were brave explorers. Marquette's mission, however, was to minister to the French soldiers on the expedition and to bring Christianity to the Indians in the Midwest.

Marquette made two trips to southern Lake Michigan. The first, in 1673, was with Louis Joliet and perhaps thirty French soldiers. The second started in the fall of 1674. On that trip he spent the winter in what is now Chicago's South Side. According to author Ulrich Danckers, the following spring, feeling very ill, Marquette most likely took a route

along Hickory and Thorn Creeks
to the Calumet River and on to
Lake Michigan.[1]

This 1675 route was possible
because Calumet Area rivers
were deeper and wider in the seventeenth century than they are
today. And the high spring level
of these creeks would have made
this trip much easier still than at
other times of the year. Historian
Powell Moore noted that "it is
even probable that they camped
at the mouth of the Grand Calumet and that Mass was celebrated
there, as so many believe."[2]

Jacques Marquette (1637–1675).
Michigan City Public Library.

There are no eyewitness documents that describe Marquette's travels
through Northwest Indiana, but since it is known that his small group
paddled an average of thirteen miles each day, it is quite possible that
they made three stops in what is now Indiana while paddling along the
south shore of the lake. And since the three streams that flow into Lake
Michigan are about thirteen miles apart, it can also be assumed that, as
suggested by author George Brennan, those stops would have been at
the mouths of the Calumet River where Marquette Park is today; Fort
or Dunes Creek, now at Dunes State Park; and Trail Creek at today's
Michigan City.[3]

Today Marquette and his voyage are honored and remembered in
several area locations. These include Marquette High School, Marquette
Mall, and Marquette Spring in Michigan City, Marquette Street in Lake
Station, and Marquette School and Park in Gary.

THE MARQUETTE PLAN: A VISION FOR LAKESHORE REINVESTMENT

In the years since 1830 when development first started on the Indiana
lakeshore, there had not been a comprehensive plan for the best use of
the full forty-six miles of shoreline until the creation of the Marquette

Plan. A Marquette planning project, named after Jacques Marquette, was first introduced by US Congressman Pete Visclosky in 1985. One of its original goals was to open up three-quarters of the Indiana shoreline for public use.

Ahead of its time, the Marquette project didn't capture the interest of the area's business, environmental, and governmental leaders until the early 2000s. By that time, the Northwest Indiana Quality of Life Council had been formed, and it provided the needed organization to begin the process.

In 2003, the lakefront cities of Gary, Hammond, East Chicago, Whiting, and Portage agreed to a memorandum of understanding to pursue the development of a Marquette Plan; they all contributed funds to match grant funds so that the plan could be developed. A working group comprising representatives of each of the five cities, the Indiana Department of Natural Resources, and consultants agreed that the plan should establish a balance between "nature and industry, between public access and privacy, between old jobs and new economies, between redevelopment and restoration and between heritage and a new way of life."[4]

The plan was created through an extensive public process that included town hall meetings in each community, interviews and input from more than one hundred stakeholders, and review meetings with industry. The planning groups decided that the plan should showcase the Calumet Area's heritage, increase public access to the lakefront, increase opportunities for recreation, protect and enhance natural landscapes and waterways, spur intergovernmental cooperation, and welcome residents and visitors to the lakefront.

The plan, which focused on the lakefront from the Illinois state line to the Burns Waterway, was approved by the five participating cities and published in January 2005. Three guiding principles agreed to were to have 75 percent of the lakeshore open for public use, to establish a minimum setback of two hundred feet from the water's edge for new development, and to create a continuous biking and walking trail along the shore through Northwest Indiana.

In the meantime, in 2005, the Regional Development Authority, a development agency covering Lake and Porter Counties, was created by the Indiana General Assembly, which gave the agency four tasks: to

use the Marquette Plan as a guide while helping redevelop the shoreline; to support expansion of the Gary/Chicago International Airport; to support passenger rail and bus transportation within the region; and to boost Northwest Indiana's economic development.[5]

One of the early suggestions of the first planning group was to extend the plan boundaries eastward to the Michigan state line so that all of Porter County and LaPorte County's lakefront would be included. That segment of the Calumet Area was planned in the project's Phase II.

Task forces and focus groups from the eastern lakefront communities then met, studied, and debated, and in September 2008 they finalized their vision for Lake Michigan shoreline reinvestment. The Phase II plan was approved and published that year and has guided actions along the eastern shoreline since then.

Each community as well as the state and national parks in the lakeshore area are responsible for implementing the projects within their own borders that were recommended by the plan. Funds from the Regional Development Authority have been used to fuel the work of various aspects of both phases of the Marquette Plan, providing the local funds needed to leverage grants from the federal government and private foundations and environmental groups.

Regional Marquette Plan projects, defined as those that impact the entire Indiana lakefront, include the following:

- The Marquette Greenway, an uninterrupted trail across Northwest Indiana that connects parks, beaches, museums, and historical buildings along the lakefront.[6]
- The Lake Michigan Water Trail, a route that includes designated places along the shore where paddlers can rest, or enter and leave the water. When finished around the rest of the lake, it will be "the longest continuous-loop water trail in the world."[7]
- Watershed planning and implementation, in order to ensure that clean, healthy water enters Lake Michigan from its tributaries. This is an essential part of the Marquette Plan. Examples of these projects are discussed in chapter 10.

Facing: Statue of Father Marquette at the entrance to Gary's Marquette Park.

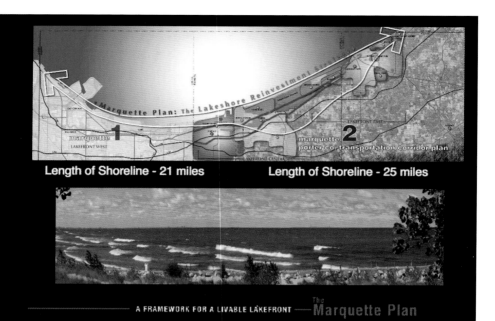

Length of Shoreline - 21 miles   **Length of Shoreline - 25 miles**

A FRAMEWORK FOR A LIVABLE LAKEFRONT ——— The **Marquette Plan**

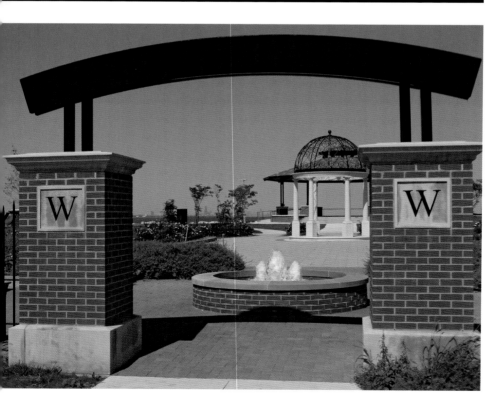

- The Calumet National Heritage Area. This is a proposal by the Calumet Heritage Partnership to have the south rim of Lake Michigan, from Chicago's Lake Calumet area to the Indiana-Michigan state line, designated by Congress as a National Heritage Area. This type of designated area can assist with efforts toward conservation of important natural resources, historic preservation, recreation, education, and tourism.

Local already-completed projects include the Hammond Lakefront Park Beach and Bird Sanctuary, Wolf Lake Pavilion, Forsythe Park, Whiting Lakefront Park, Jeorse Park improvements, Roxana Marsh, sections of the Grand Calumet River restoration, Marquette Park rejuvenation and restoration, the Portage Lakefront and Riverwalk, the Indiana Highway 49 Lakeshore Gateway Corridor, the Orchard Pedestrian Way Trail, the Sea Lamprey Barrier, Charles Weston Gateway Park improvements, Michigan Boulevard improvements, and the Hansen Park ADA Route and Kayak Launch System.

Facing: The Marquette Plan (*top*) and Whiting Lakefront Park (*bottom*).

Harry Eenigenburg (1859–1943). *Marie Min.*

Natural Resources of the Calumet Area

3

It really was the most beautiful country in the Middle West with all its streams, high sand ridges, big trees, meadows and marshes.

Harry Eenigenburg, *The Settlement of the Calumet Region*

THE CALUMET AREA LANDSCAPE FOUND BY THE ORIGINAL SET-tlers was a fascinating wilderness composed of sand dunes and ridges, quiet rivers with crystal-clear water and extensive floodplains, dense forests, prairies, savannas, and marshes, and one "great" and many small lakes. These many landscapes then provided a home for great diversity in plant and animal life. While the Calumet Area has never had natural deposits of gold, silver, petroleum, or iron ore, it has had plenty of other natural resources highly valued by pioneer and modern entrepreneurs. These included timber, sand, clay, water, ice (in the winter), wildlife, transportation routes, and rich farmland. Harry Eenigenburg, whose family was one of the first in the Oak Glen (Lansing), Illinois, area, wrote that before the area was changed into residential areas and farmland, "the rivers and creeks, the sand dunes, the big trees, [and] the canebrake marshes provided for the settlers a choice home, for everything they needed nature had provided in great abundance."[1]

TIMBER

As two of the first needs of the incoming pioneers to this area were housing and fuel, Calumet Area trees were quickly felled to provide both, and among the first industries to be established in the region were sawmills. The lumber was used for local building purposes, but it was also exported to Chicago and other areas where trees were no longer found.

Hammond, Whiting, Hobart, and Michigan City all became centers of the Calumet Area timber trade. Trees were everywhere except for a few natural prairies.[2] Pine Township in eastern Porter County is even named for the thousands of native pine trees that used to grow there.

Within the first few years of settlement, the trees started to disappear. Farmers cleared level areas so that they could farm. Railroad men used trees for crossties and locomotive fuel. Everyone needed wood for shelter, for cooking, and for heat in the winter. As early as 1847, just ten years after land sales began in Lake County, pioneer Solon Robinson, the founder of Crown Point, noted that several of the county's sand ridges, once covered with pine and cedar trees, had been "nearly all stript off to build up Chicago."[3] Areas to the east were affected the same way. LaPorte County historian Jasper Packard lamented the loss of the once extensive forests of "beautiful white pine." That species, he wrote, "has almost wholly disappeared, being cut off for lumber."[4]

Harry Eenigenburg reported that the years 1870 and 1871 were the driest "on record. Scarcely any rain in two years."[5] He noted that the Cady Marsh area, which had recently been drained, caught fire on October 8, 1871, the day before the Great Chicago Fire started. Earlier that summer, the ground was so dry in the Whiting area that the Forsythe family buried their large collection of books in the sand to protect them.[6] But, of course, fires did occur. With so much timber destroyed and so much construction required after these fires, the demand for and the price of lumber skyrocketed. Historians Timothy Ball, George Garard, and Powell Moore all reported that it wasn't uncommon for thieves to search for large trees in Lake and Porter Counties, cut them down at night, and transport them to Chicago.

SAND

In addition to guiding the restoration of the parts of Chicago that were destroyed by the fire, the city administration decided to expand eastward. Rubble from the fire may have started this process when much of it was simply dumped into the lake. It was later decided to deliberately obtain sand and dump it into the lake to create a new open lakefront, including Grant Park. Many Chicago trucking companies purchased duneland in Lake and Porter Counties and commenced mining the In-

A steam shovel taking sand from a dune in Miller and dumping it into waiting railroad cars. *Calumet Regional Archives.*

diana sand and hauling it to the city. The 1893 Chicago World's Fair required even more sand.

Much of the ground in the northern portions of the Calumet Area is sand, a natural resource that was easily mined, and easily transported if there was a railroad nearby. It was in great demand not only by the city of Chicago, but by numerous midwestern industries as well.

By the 1890s, sand had become to Northwest Indiana what coal was to West Virginia. At the Miller and Dune Park (today Portage) railroad stations, located right next to the Dunes, more time was spent and money made by the shipping of sand than of any other commodity. The same was true at Tolleston, which was miles away from the Dunes but centered on the High Tolleston Shoreline, that four-thousand-year-old former shoreline of Lake Michigan.

The numbers are hard to imagine. According to historian Powell Moore, in 1897 about 50,000 railroad cars full of sand were loaded in the Tolleston area and shipped out. In 1899 150,000 cars of sand were shipped out of Miller for the Santa Fe Railroad. The mining of such huge amounts of sand made an equally huge impact on the landscape. By about 1920, in an area near today's Port of Indiana, an entire square mile of sand dune was mined and sent to Chicago.[7] Michigan City historian E. D. Daniels noted in 1904, "There is practically an inexhaustible amount of sand, yet under this steady demand some of the sand dunes on our lake shore are diminishing very noticeably. Hoosier Slide [in Michigan City] is not nearly as high and large as it used to be."[8]

Large-scale mining went on for decades. As late as 1952, the year that the Save the Dunes Council was formed, the Indiana Geological Survey reported that each day five thousand tons of Calumet Area sand were still being removed. The authors of the report believed this to be quite positive. Although they admitted that areas "once comprising some of the most spectacular and picturesque dunes" had already been mined out, they maintained that there were problems associated with building on dunes, and so "the average householder probably will be better satisfied to live on the more level ground of the worked-out sand area."[9] In spite of the fact that, at that time, the mining industry was still strong, the authors predicted that the remaining sand dunes in Northwest Indiana could nicely fill the needs of progress for another fifty to one hundred years.

The Hoosier Slide, a large sand dune, used to sit right at the mouth of Trail Creek in Michigan City. Although it was never official, local folks said that at two hundred feet it was Indiana's tallest dune. Its unusual name was given to it because of its occasional avalanches of sand. Mining of that sand dune began around 1890. Just ten years later, in his history of Northwest Indiana, Timothy Ball reported that even though "immense quantities" of sand had already been hauled away and shipped to Chicago, it was "a huge mass yet."[10]

Because of an impurity in this dune's sand, glass made from it had a strange bluish tint. Thus for several years, companies that wanted to make crystal-clear glass were not interested in it. That changed when the Ball Company realized that tinted glass made foods stored in jars last longer. Thus tons of sand from this suddenly popular dune were sent to glass factories in north-central Indiana. Other sands were hauled off to beautify sandless resort lakes and to build up railroad rights of way. By

Two Calumet Area natural resources: sand (the Hoosier Slide in Michigan City) and water (Trail Creek). Circa 1910. *Old Lighthouse Museum.*

the mid-1920s all the sand from the Hoosier Slide had been mined. The flat land that remained was then sold to the Northern Indiana Public Service Company (NIPSCO), which built its generating station right there, where once stood what was said to be the greatest sand dune in the country.[11]

With few exceptions, the dune-and-swale topography that characterized the landscape of Whiting, East Chicago, and the northern parts of Hammond and Gary was smoothed off before city streets were constructed and lots sold. One beautiful exception was two city blocks of a mile-long swale in front of Gary's Horace Mann School. That swale (lagoon) was saved and reshaped as a reflecting pool. Schoolchildren in the 1930s planted wetland plants along its edge, and for about thirty years, residents in the area skated on it every winter. The lagoon, complete with

Clay pit at Hobart, circa 1895. Note the size of the horses and men in the background. *Blatchley*, Geology of Lake and Porter Counties, *128.*

swans, graced the school from when it was built in 1926 until 1959 when it was covered over with a parking lot.[12]

CLAY

The moraines and lake bottomlands of the Calumet Area contain an abundance of clay, and folks realized early on that much of it was good enough to make bricks. Small brickyards were soon found throughout the morainal area. Large yards were located in Munster, Hobart, and Porter, all of them alongside railroad lines so that the bricks could be shipped to distant cities. By 1920, however, most of the brick factories had ceased operations. In 1987, the American Brick Company in Munster, the last yard still operating in the Calumet Area, stung by complaints about its smoke and smells, closed.

Although there are no brickmaking plants still operating in Northwest Indiana, many of their excavation areas still dot the landscape. Maynard Lake in Munster and Hidden Lake in Hobart are former brickyard excavation pits. So is the low-lying area around Yost School in Porter. Many clay pits have been used as city dumps or, more recently, repurposed as sanitary landfills.

Clay was also used extensively to raise roadbeds. To save transportation costs, this was often done by excavating clay from a location adjacent to the highway, resulting in a deep hole next to the raised roadway. Many of these "borrow pits" have been flooded by groundwater and today can be seen as ponds or small lakes. Some have been utilized as recreational areas and are now surrounded by resorts or parks. Lake Minnehaha at Jellystone Park along Interstate 94 in Portage is a former borrow pit.

With so much Calumet Area sand and clay being mined, there was bound to be at least one strange, interesting twist to the story. As noted above, in the latter part of the nineteenth century, a large amount of sand from Miller (a Gary neighborhood) was mined and shipped to Chicago. Then about twenty years later, good black dirt was removed from the Munster brickyards so that the clay underneath could be mined. The brick company found a grateful customer in the Gary Land Company, which needed black dirt to cover the sandy yards in the new neighborhoods it was building. Then in the 1990s, Chicago's Museum of Science and Industry decided to move its parking lot underground. In doing so, the museum excavated a huge amount of clay that was then sent to Munster to be used as cover for the town's sanitary landfill—a landfill at the same brickyard where the topsoil had been excavated and sent to Gary some eighty years before.[13]

WATER

A most plentiful natural resource in the Calumet Area, water is becoming more valuable each year. Plentiful rainfall has for millennia watered the forests, savannas, and prairies of the area and provided water for drinking and for crops for the Miami and Potawatomi Indians, the area's settlers, and residents today.

Running water in streams and rivers provided transportation routes not just for the Indians, but also for French Canadian voyageurs, and early settlers. It now provides pleasurable recreation, not to mention great exercise, for paddlers.

In 1837, water's abundance was reflected in the decision to name Indiana's far northwest county Lake County. Beginning with the pioneer era, plentiful water encouraged settlement and farming. By the middle of the nineteenth century water was powering sawmills and gristmills. By the end of that century, steel mills were locating near Lake Michigan in order to use its water for both transportation and cooling.

Water is probably the most important natural resource in the Calumet Area—one that it shares with hundreds of other Great Lakes communities. Water in Lake Michigan serves our transportation businesses. The Port of Indiana at Burns Harbor is the busiest port in the Midwest. Still growing, in 2014 it handled more tonnage than in any other year since it first opened in 1970.

The water in the area's many streams and lakes serves as a home for fish and wildlife. Lake Michigan offers recreation along much of its forty-five-mile Indiana shoreline. Water in Lake Michigan is also the area's primary source of drinking water, while water in the ground serves as a major source of irrigation and drinking water for those who are not able to get Lake Michigan water. Water from both sources is used for cooking, washing, and sanitation, for recreation in swimming and wading pools, and by fire departments to quench fires. Water in public fountains delights the public.

Calumet Area water is plentiful and renewable, but not limitless. It is easily taken for granted and easily polluted. However, it is now protected by the Great Lakes Compact, a binational treaty with Canada, and, for its own protection, is not for sale to communities outside its natural drainage basin.

All of the cities and most of the towns in the Calumet Area established waterworks early in their development. Michigan City organized a water department in 1875. Hammond's first waterworks in 1887 was a private company that pumped groundwater from several rather shallow wells. Then in 1903, the city built a water-intake crib in Lake Michigan and finally began using Lake Michigan water, but it was thirty years before it built a filtration plant. Up to that time the city residents were provided with raw lake water that was frequently unpalatable because of pollution and what the water department called varying shore conditions.

The fanciful fountain at the entrance to Munster's Center for Visual and Performing Arts. *Center for Visual and Performing Arts.*

When Standard Oil started building its refinery in Whiting in 1889, its first project was the construction of a waterworks with a twenty-inch-diameter pipeline out to Lake Michigan. A short time later a tunnel was dug, connecting to a crib a half mile out into the lake. Soon Standard Oil was providing water to the whole city from its lines.[14]

Gary installed its water lines as the city streets and alleys were being constructed in 1906–1907.[15] By 1915 there were forty-one miles of water mains under the city thoroughfares. The main line from the lake was fifteen thousand feet long and seventy-two inches in diameter. The importance of plentiful water at that time is reflected in the handsome pumping station and iconic water tower at the city's Jefferson Park (now Borman Square). The 110-year-old tower is still in use.

Gary's first water pumping station and the octagonal water tower
in Jefferson Park, 1913. *Calumet Regional Archives.*

The towns and cities too far from the lake in the early days to take
advantage of its water still had plenty of water for their populations, but
since glacial subsoil is not always a good aquifer, it often took a while to
find the best place to sink their pipes. Furthermore, well water sometimes either smelled or tasted bad. In some areas well water had so much
iron in it that bathtubs and shower stalls often turned brown from the
iron oxide residues left behind when the water evaporated.

Hobart established both its water and electric works in 1897, and although the city lies next to Lake George, water was pumped up from
the ground "in adequate amount and perfect purity," as was said at the
time.[16] Nevertheless, as the city grew, citizens were pleased to be provided Lake Michigan water by the Gary-Hobart Water Corporation.

Munster's 1923 water tower on Calumet Avenue north of the town hall. *Munster Historical Society.*

Munster's story is similar to that of other suburban communities. Until 1923, all the residents of the town had their own wells and pumps. That year, the town established a public waterworks, drilled a well, and constructed a water tower on the ridge just behind the town hall. A second well was drilled in 1928 and a third in 1932. In 1938, with a growing population, an adequate groundwater supply was more difficult to extract and so the town got a $20,000 grant from the Depression-era Public Works Administration and built a connector line to Hammond in order to bring plentiful and better-tasting Lake Michigan water to town. Since then the town has built several additional water tower tanks to accommodate the citizenry.

Crown Point's new water storage tank east of I-65.

Today, the cities and towns along the South Shore of Lake Michigan, as far south as Crown Point, have built large water storage tanks to serve the hundreds of thousands of residents, businesses, and industries in the area. As long as communities are within the natural Lake Michigan drainage basin, they are welcome to as much of Lake Michigan's waters as they need—because their treated wastewater is returned to the lake.

ICE

Cold Calumet Area winters and a large number of small lakes and slow-moving rivers meant that there was plenty of ice in the winter. Once the railroads had crisscrossed the region, investors realized that they could make a handsome profit by harvesting the ice, storing it in insulated buildings, and when warmer weather arrived, shipping it to midwestern locations that didn't have the "hydro-advantages" found in Northwest Indiana. The ice was beneficial to businesses in other ways, as well: for instance, one of the reasons that George Hammond's meatpacking business moved its operations to the Calumet Area was because there would be enough ice for its insulated rail cars.

Before refrigeration, ice was in great demand every summer. Early on, many farmers and their sons would break up local river ice and haul it back to the farm where it would be buried in pits under the barn and covered with sawdust to keep it from melting too fast. Such stores of ice would then be used when warm weather arrived in the spring and summer.

Powell Moore reported that John Vater and Heinrich Eggers "pioneered" the shipping of ice from Berry Lake in the Whiting area to Chicago.[17] But most of the ice harvested was done by companies such as the Knickerbocker Ice Company, which built huge storage houses alongside the Calumet Area's small natural lakes and employed hundreds of men every winter to cut and haul the ice. Getting ice from Lake Michigan was dangerous then, as it would be now, so most of the ice was harvested from Wolf Lake, Cedar Lake, Long Lake, and the lakes north of Valparaiso and LaPorte. According to *Industrial Chicago: The Building Interests*, published in 1891, "the Knickerbocker Ice Company's mammoth icehouses on Wolf Lake [were] the largest in the world."[18]

Harvesting ice on Wolf Lake in Hammond. *Hammond Public Library.*

Harvesting ice on Long Lake in Miller. *Lake Street Gallery.*

Lake George Ice Company's ice wagon. Note the small window near the front of the cab. *Whiting-Robertsdale Historical Society.*

Ice was then shipped to locations all over the Midwest, but the largest sales area was Chicago, where Knickerbocker and E. A. Shedd & Company had their headquarters. Once the ice was shipped to the city, it was put on wagons and ice sellers would ply the neighborhoods, selling ice to households for their iceboxes.

When the weather was warmer both Knickerbocker and Shedd employees were busy mining sand. The advent of electric home and business refrigerators in the 1920s quickly put an end to the once thriving ice business.

WILDLIFE

It was the area's wildlife—fish, wild birds, and mammals—that helped feed American Indians for thousands of years. That same wildlife brought French voyageurs to the Calumet Area beginning in the late seventeenth century. Beaver, mink, and muskrat pelts were highly sought before and for several decades after the American Revolution. Lake Michigan teemed with fish, which not only provided food for the American Indians but also allowed a thriving fishing industry on the lake for many decades.

Gustav Leafgren (*right*) and a fellow iceman with the Consumers
Company deliver ice to residents of South Chicago. *Selma Swanson.*

Even after the development of cities in this area, the forests and wetlands provided food for local farm families and sport hunting for the
wealthy. In the late nineteenth century Lake County alone had several
hunting clubs, known as gun clubs, located in the wet marshy areas both
north and south of the Valparaiso Moraine.

Before the founding of the city of Gary, fish and game were plentiful
in and along the Grand and Little Calumet Rivers. An estimated thirty
thousand muskrats were trapped each year, their skins could be sold for
five or ten cents each,[19] and one man could shoot fifty to one hundred
ducks in just one day.

The Tolleston Gun Club of Chicago (originally the Tolleston Shoot-
ing Club) was established in 1871 and purchased about two thousand
acres of marshland that paralleled the Little Calumet River south of the
little village of Tolleston. The club's grounds served as a recreation and
hunting resort for one hundred wealthy and influential Chicagoans such

The Miller or Calumet Gun Club building, 1903. An open porch on the north side of the building allowed convenient fishing in the Grand Calumet River. *Jean Buckley.*

as its treasurer, C. D. Peacock; John W. Gates, president of Illinois Steel; and Marshall Field, department store owner and a founding director of U.S. Steel. Tolleston was close enough to Chicago that club members could come out for weekend shoots. Frederick Howe, president of the club from 1873 to 1896, noted, "We didn't have time to take a week away from our business, but we could easily run out to Tolleston to stay over Saturday and Sunday."[20]

Built on the grounds was a large two-story clubhouse with a reception room, a ladies' parlor with adjoining bedrooms, a dining room, a kitchen, and room for forty beds. Nearby was a large barn, an icehouse, dog kennels, and a boathouse connected to the river by a small canal dredged for the club.

Within a short time, the club members' sporting culture conflicted with the local farmers. The organization fenced much of the open river-bottom areas that had long been used in common by the farmers. While

The Tolleston Gun Club clubhouse. Much larger and finer
than the Calumet Gun Club building, it showcased the
economic power of its members. *Calumet Regional Archives.*

the farmers trapped or shot animals for their own dinners or to sell for
some spending money, the club members, of course, hunted for sport.
(It was a common practice to shoot ducks before breakfast—the club
record stood at 139 ducks bagged all in one morning.)[21] The club's game
wardens considered farm boys who wandered onto the Tolleston Club
property as fair game. Family historian Peter J. Schoon later recalled
that the club members "weren't interested in running trap lines because
that's hard work; you have to get up early to do that."[22] The local residents resented the wealthy intruders from Chicago who permitted and
enjoyed the killing of game for sport but "denied the poor man a brace
of ducks for his table."[23] Although a fatality in 1894 didn't seem to have
much effect on the club, not so an altercation in 1897 that involved more
than twenty men.

According to Sam Woods, the son of early Lake County settlers, the
Tolleston Club falsely claimed that it owned a lot more land than it actually had. If any of the local farmers hunted near the club grounds, the
club wardens would insist that they leave. If any of the farmers then
refused, "they would be very roughly handled."[24] This went on, Woods

wrote, until the local folk grew desperate, and on January 19, 1897, a group of about twenty young farm lads went on a muskrat expedition and "sallied forth on the raging Calumet."[25] Resident Bill Lohman, whose descendants still live in the area, was reported to have claimed that they were not trespassing on club grounds but were actually on land he was renting, and that they had permission to hunt there. In any case the watchmen saw them and ordered them to leave. The lads evidently refused. Both sides later maintained that the other fired the first shots; the locals also claimed that they were shot when they were leaving the grounds—which Woods noted must have been true, "for all of them were shot in the back."[26]

Chicago newspapers proclaimed, "Deadly Conflict in Tolleston," "Club Guards and Farmer Boys Use Their Guns," and "Citizens Beaten and Maimed." The *Chicago Dispatch* wrote of "farmers of Tolleston marsh . . . driven from their own lands by Bullies hired to protect the Preserve of the Rich." The papers compared the club members to "barons of old" and claimed that the Tolleston Club "habitually defies the laws of Indiana" and that the "title to land is disputed and legal proceedings will follow."[27] Luckily, the wounded men recovered. But after all this, public opinion turned against the club and it disbanded after the turn of the century.

The seal of the State of Indiana features the image of a pioneer woodsman chopping a tree and an American bison jumping a log. Bison were seen in both Indiana and Illinois in the early days of settlement, but none since 1832. None are known to have been seen near Lake Michigan. Large mammals once seen in Northwest Indiana included elk, panthers, and bears, but as their habitat was changed from forest to settlements and farmland their populations decreased, and they were gone by about 1850.[28] However, in 2015 a wild bear did, for at least a few weeks, return to the state, crossing from Michigan into the Michigan City area. Bald eagles and peregrine falcons, once plentiful along the Lake Michigan shoreline, were gone by 1897. The hated timber wolf was hunted to extinction in this area by 1908. The last passenger pigeon was seen in 1902.[29] As a result of efforts by the Indiana Department of Natural Resources, the eagle and the falcon have been reestablished in this area. In his book

Birds of the Indiana Dunes, Kenneth Brock notes that most sightings have been along the lakefront; most of the birds were seen in spring and fall on their migration journeys, but some are now nesting there.[30]

TRANSPORTATION ROUTES: "LOCATION, LOCATION, LOCATION"

Lake Michigan affected Calumet Area transportation routes in two ways: it provided north–south water transport and prevented east–west land transport. The former was used by seventeenth-century French explorers and is still heavily used by industrial shipping. The latter caused east–west Indian trails to go around the southern shore of the lake and still causes modern highways to do the same. Thus the transportation industry has played a major part in the Calumet Area's history.

Beginning with the Sauk and Calumet Beach Indian Trails, which became stagecoach routes, which later then became state highways, east–west travelers moved through Northwest Indiana. These transportation routes were served by scores of inns to lodge weary travelers. Indeed, the hotel industry is one of the oldest types of businesses in the area.

All the rail lines in the nineteenth century connecting the East Coast with Chicago passed through Northwest Indiana. This part of the country still has one of the highest concentrations of rail lines in the United States.

With the advent of the automobile, major highways began to extend across the area as well. The nation's first transcontinental highway, the Lincoln Highway, followed much of the old Glenwood Shoreline of Lake Michigan as it crossed Lake County.

US Routes 6, 12, 20, 30, and 41 and Interstates 80, 90, 94, and 65 make for one of the country's busiest highway systems. Today service stations, hotels, and truck stops continue to serve travelers as well as provide employment for local residents. They can all be expected to continue to do so well into the future.

FARMLAND

The gift of the glaciers has been an extremely thick layer of glacial till (sediment) spread out through most of the upper Midwest. Indiana is fortunate in this respect because it is far enough north to have been

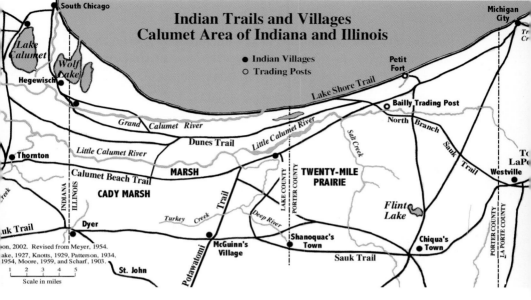

Indian trails and villages in the Calumet Area of Indiana and Illinois. *Schoon*, Calumet Beginnings, *50*.

Major Calumet Area rail lines in 1910. *Schoon*, Calumet Beginnings, *86*.

nearly completely covered by till and far enough south to have a long enough growing season for most crops. Although the Calumet Area is often thought of as being primarily industrial, much of it was originally agricultural. Before the 1950s a considerable part of south Hammond and midtown Gary and much of the land south of the Dunes in the eastern Calumet Area were also agricultural.

C. P. Schoon, a man outstanding in his field. This crop of flowering onions, at what is now the corner of Ridge Road and Calumet Avenue in Munster, was grown to harvest the seeds to sell to other Illiana-area farmers who then grew onion sets. By the middle of the twentieth century, the Calumet Area was producing more than 1.5 *billion* onion sets each year. That was half the country's total! *Schoon family collection.*

Industrialization of the Lakefront

4

IN THE NINETEENTH AND EARLY TWENTIETH CENTURIES, Americans were awed by the power of the Industrial Revolution. Chicago's World's Fairs in both 1893 and 1933 celebrated it. While people flocked to Chicago to visit the fairs, people were also flocking to the cities looking for jobs and, through these jobs, prosperity.

The Calumet Area was in the center of it all. Not only did Lake Michigan provide plentiful drinking water, but it could satisfy all the water requirements for the steelmaking industry, with its need for vast amounts of water for cooling. The lake also provided (and still provides) a water route to the huge ore fields of Minnesota. By 1890 Northwest Indiana was already served by thirteen major rail lines. By 1910 that number would grow to twenty, all of which could serve area business and industry. By 1953 the bistate Calumet Area steel mills were producing 20.6 million tons of steel a year and had surpassed Pittsburgh as the most productive steelmaking region in the country.

The first industries to locate in Northwest Indiana generally got their land at very affordable prices. East Chicago even offered free land to Inland Steel and promised that there would be sufficient rail lines to serve the new company. Once the steel companies were established, then other companies that used steel wanted to be nearby.

From the end of World War II until the 1970s, steel continued to be one of the greater Chicago area's strongest industries. At that time the United States was making more than half of the world's steel, and 20 percent of that was produced in Calumet Area mills. Many of the large plants established in the early part of the century continued to make

Otto Brennemann's 1926 South Shore Line poster celebrating the
industrial might of the Calumet Region. *Calumet Regional Archives.*

large amounts of steel. In the early 1960s, the Calumet Area became home to two of the country's first basic oxygen furnaces, which were more efficient and less costly to operate than the older open hearths. By the time Midwest and Bethlehem Steel built their new plants at Burns Harbor, the Calumet Area was already the center of the US steel industry. By 1990, the steel mills in South Chicago had closed, and yet the mills along the lakeshore from East Chicago to Burns Harbor alone were making about a quarter of all the steel produced in the country.[1]

In Northwest Indiana, it was the new railroads that brought with them the era of manufacturing. The earliest railroad arrived first in La-Porte County, so it is not surprising that the establishment of large factories happened first in that county.

LAPORTE COUNTY

Heavy industrialization of the Calumet Area began in Michigan City soon after the arrival in 1850 of the first railroad to that area, the Michigan Central. Enterprising folks at that time realized that as the rail industry grew, so would the demand for railroad cars.

Michigan City was the first city in Northwest Indiana to be platted, and for decades it was the largest Indiana city on the lakeshore. Major Isaac Elston, who founded the city, believed that its location on the southeast side of Lake Michigan would mean it would soon grow larger than Chicago.

The industrialization of Michigan City began in 1850 with the arrival of the Michigan Central Railroad. Two years later the Michigan Central established its repair shops in the city, and that same year a small rail-car manufacturing plant was started. By 1860 its sixty-man workforce built a couple of freight cars every month. In 1871 the company was incorporated as the Haskell & Barker Car Company. Eight years later, five hundred men were making one thousand freight cars a year, and by 1907 the company employed more than 3,500 men and was making fifteen thousand freight and passenger cars a year. By this time Haskell & Barker claimed to be the largest freight car company in the world.[2] It had more employees than any other company in the state, working in forty-five buildings spread over one thousand acres in the city. To power the plant, the company burned seventy-five thousand tons of coal each year. The

Haskell & Barker Car Company, 1916 postcard. *Old Lighthouse Museum.*

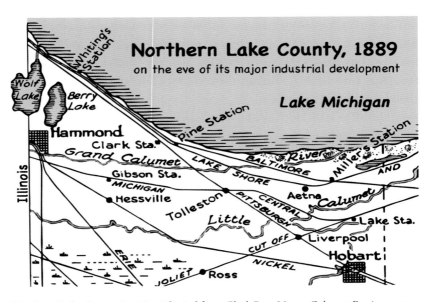

Northern Lake County in 1889. Adapted from Clark Ray. *Moore,* Calumet Region, *179.*

plant put massive amounts of black smoke into the air every day, but at the time, the smoke was an outward sign that there was employment not just for factory workers, but for many, if not most, of the construction workers, grocers, teachers, and others in the area.

In 1922 the Haskell & Barker works were purchased by Chicago's Pullman Company, and Edward F. Cary, who then owned H&B, became president of Pullman. Later the Michigan City works were known as the Haskell & Barker Shops of Pullman-Standard.[3] The plant continued operations until it closed for good in 1971.[4] The Lighthouse Place Outlet shopping center now sits where the country's largest freight car plant once stood. The largest employer in Michigan City today is Blue Chip Casino.

LAKE COUNTY

As late as 1889, the Lake County lakeshore was largely undeveloped, still a wilderness, uneven, inhospitable, wet, and sandy. It had been abandoned much earlier by even the stagecoach route that had once followed the beach to Chicago. A few rail depots near the state line and the Carr family's fishing operation in Miller were the only signs of civilization. Hammond and Hobart, the only nearby cities, were both located south of the lakefront area. They also depended upon the railroads for employment.

In 1870 the US Congress initiated development along the Calumet Rivers when it provided funds for a new harbor in Illinois at the mouth of the Calumet River in South Chicago. The river was straightened and dredged and soon provided sites for heavy industry such as Brown Iron & Steel (later Wisconsin Steel), the North Chicago Rolling Mill (U.S. Steel's South Works), Iroquois, and Republic Steel.

Standard Oil came to Indiana nineteen years later. In 1901, just southeast of Standard Oil's Whiting refinery, and while Inland Steel was building its first Indiana mill, a new harbor and canal were being dredged where no harbor—or even a river—had been before. Those two efforts resulted in what would soon become Indiana's busiest harbor and its first open-hearth steel mill. Construction of the area's first cement plant began at Buffington Harbor in 1903. In 1906, to the east in Gary, the United States Steel Company began work on its mill, soon to be the largest integrated steel mill in the world. Finally, in 1914 back in East Chicago, Mark Manufacturing (later LTV Steel) opened its mill across the canal from Inland's plant.

Urban and industrial development occurred simultaneously. East Chicago was incorporated in 1889, Whiting in 1895, and Gary in 1906. All grew quickly with large working-class neighborhoods. Gary soon became Indiana's second-largest city. In less than twenty-five years, it had a population of more than one hundred thousand.

The story of Calumet Area industrial development is rather spectacular and cannot be told without using superlatives. Lake County alone had the world's largest refinery, largest cement plant, and largest integrated steel mill. Its industries employed thousands of men and women, provided petroleum, steel, and steel products for two world wars, and then afterward played their part in the postwar boom of the mid-twentieth century. This rapid development of the lakeshore area all began when John D. Rockefeller decided to build his largest refinery in Indiana.

WHITING AND STANDARD OIL

In the 1880s, the products of the oil industry were fuel oil, kerosene for cooking and lighting, lubricating oils for locomotives and industry, and grease for horse-drawn wagons and carriages. In 1885 when a large oil field was found near Lima in western Ohio, about sixty miles southeast of Fort Wayne, the Standard Oil Company invested heavily in its oil fields. Unfortunately, that petroleum contained sulfur and smelled like rotten eggs. Kerosene made from it smoked badly and gave off a foul odor. Luckily in 1887 Herman Frasch discovered that refining oil in the presence of copper oxide removed the sulfur and left the oil odorless. Standard Oil hired Frasch, bought his patent, and decided to build a refinery in or near Chicago, the largest city in the Midwest. At that time there was no oil refinery in the United States west of Ohio. Standard Oil thus became the first major industry to be built along Indiana's Lake Michigan shoreline.

The location near Whiting Station was chosen because the land was available and fairly cheap, there were plenty of railroad lines nearby, and the taxes were lower than Chicago's but it was still close to the city. Construction of what would become the world's largest oil refinery began in 1889. The first shipment of kerosene, contained in 125 tank cars, left the

Northern Lake County in 1933. Adapted from Clark Ray.
Moore, Calumet Region, *between 336 and 337.*

plant on Thanksgiving Day 1890. Kerosene remained its most important product until 1910.[5]

Before Standard Oil arrived in what today is Whiting, the town was, as Powell Moore observed, "a sleepy hamlet clustering around Henry Schrage's store, its only business establishment."[6] The ground was uneven—originally composed of a series of long parallel sand ridges separated by shallow wet swales created as the Lake Michigan shoreline gradually receded over the previous 1,500 years.

Moore described the presettlement Whiting area as "one of the most uninviting areas of the region."[7] It was easy to walk or ride along a ridge, but to cross to another ridge could be very difficult. The largest swale, so large that it was often called a lake (Berry Lake), was on the east side of the town. As the community and its industry were developed, the ridges were leveled and many of the long, narrow wetlands, including Berry Lake, were drained or filled in. Even as construction of the refinery was about finished, the area was so wet that boats were often used by workers to get from one storage tank to another.

One of Standard Oil's major innovations was its establishment of a research laboratory. Perhaps its greatest accomplishment was the 1909 invention of the "Burton cracking process," which greatly increased the amount of gasoline obtained through refining and was considered by some to be among the greatest inventions of modern times. This laboratory paid great dividends in years to come as the company became the industry leader in research. By 1920 there were more than four thousand employees at the plant.

Early in the morning on August 27, 1955, an explosion broke open Fluid Hydroformer Unit 700, claimed then to be the largest in the world. Shrapnel from the unit flew into the air and a fire broke out that would eventually destroy most of the refinery's tank farm over the next couple of days. Several nearby houses were destroyed by the blast and more than 1,400 Whiting residents were evacuated. One of the worst industrial fires in North American history, it took eight days before it was quelled. Sixty-seven tanks were destroyed and 1,250,000 barrels of crude oil were consumed by the fire, which could be seen in Gary and Chicago.

The refinery was back in operation at pre-fire levels by the end of November, but so many changes have been made since the fire that an employee from the 1950s wouldn't recognize the refinery today. The modern and recently expanded refinery, now 125 years old, has double the capacity that it did in the 1950s. Lessons were learned from the fire; in particular, the problems with the refinery design that allowed the 1955 fire to spread so rapidly were corrected years ago. Today, just as in 1955, the Whiting refinery remains the largest inland refinery in the United States. Standard Oil became Amoco in the late 1950s, which merged with BP in 2000.[8]

Over the years the refinery has grown. Begun in 1889 with a 235-acre plant and processing 600 barrels of crude oil daily, it now employs 1,850 full-time employees at its 1,400-acre refinery, which now extends from Whiting into both Hammond and East Chicago. According to BP's *Facility Fact Sheet*, the Whiting plant processes up to 428,000 barrels of crude oil each day, which it receives via pipelines. From that, the refinery makes approximately nineteen million gallons of products each day, about half

STANDARD OIL CO. REFINING PLANT. NEAR EAST CHICAGO-INDIANA HARBOR, IND.—12

Standard Oil refinery, undated, colorized postcard. *Steven R. Shook.*

of it gasoline; the other half includes jet fuel and ultra-low-sulfur diesel fuel. These are shipped out through other pipelines throughout the Midwest. The plant also makes about 8 percent of all asphalt used in the United States.[9]

The refinery footprint is much different today than it was in 1955. The processing units are no longer in proximity to the storage tanks and the individual processing units are farther apart from each other. No longer are there common dikes for more than one tank. And most importantly to both the company and the citizenry, there is now a buffer zone between the refinery itself and residential neighborhoods.[10]

EAST CHICAGO: THE INLAND STEEL AND MARK MANUFACTURING COMPANIES

Unlike Whiting, where Standard Oil located before there was a city, East Chicago was planned first and industry arrived later. In 1881 the East Chicago Improvement Company purchased eight thousand acres adjacent to Lake Michigan. In 1887 a plat for East Chicago was laid out and crews started leveling the old dune-and-swale landscape. About that same time, work was started on the ship canal. Graver Tank Works, arriving in 1888, was the first industry to set up shop. The biggest prize, however, was Inland Steel, then located in Chicago Heights, Illinois.[11]

In 1901, the Lake Michigan Land Company, which then owned about 1,300 acres of land where Indiana Harbor is now, offered the Inland Steel Company 50 free acres of land adjacent to the lake if Inland would invest at least $1 million in building a steel mill there.[12] In July 1902 the first steel ingots were poured. In November the sheet and bar mills were running and Inland was employing about 1,200 workers. A galvanizing department was started in 1905. The company purchased an iron ore range in Minnesota and decided to make its own pig iron and steel. So in 1906 Inland built blast furnaces, which were required to turn iron ore into iron. Needing more land for expansion, it started filling in the lake and creating new land. In 1907 the first shipment of iron ore ever unloaded in Indiana was sent to Indiana's first-ever blast furnace.

The Mark Manufacturing Company gave East Chicago its second steel mill. The company (later Youngstown Sheet and Tube, later still LTV Steel) erected its plant in 1914 directly across the ship canal from Inland Steel. Clayton Mark, the company's owner, earlier had his Lake Forest, Illinois, home designed by well-known architect Howard Van Doren Shaw. He then commissioned Shaw to design and build an entire residential neighborhood, including parks and commercial buildings, for his East Chicago employees. Shaw created a community reminiscent of an English country village. Mark named it, of course, Marktown. The entire community was placed on the National Register of Historic Places in 1975.

Within a few years more than fifty industries were located in the city and the harbor had become Indiana's largest and busiest. Knowing the importance of having adequate transportation capabilities for its various industries, the city hyped its new harbor and canal and its multiple rail lines with the slogan, "Where rail and water meet."[13]

Inland Steel was one of the many steel mills that contributed to the war effort during World War I and then, after a brief postwar slowdown, expanded throughout the 1920s and '30s. A 375-acre expansion in the '20s extended its property an additional mile into the lake. By 1939, the Indiana Harbor plant, with four blast furnaces, employed more than nine thousand employees.

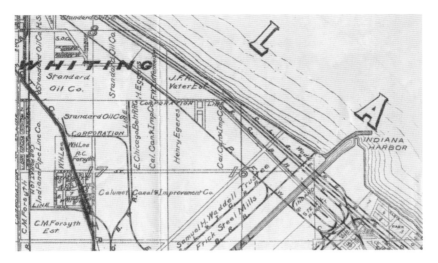

The industrial area of Whiting and East Chicago in 1906, featuring the Indiana Harbor Ship Canal and the land then owned by Standard Oil and Inland Steel. Note that Inland had already begun expanding into the lake. *Jones, Northern Portions of Lake County. Courtesy of the late Norman Tufford.*

Illustration from a 1911 Inland Steel Company sales brochure. Smoke was deliberately shown emanating from multiple stacks because that indicated that work was in progress. *Calumet Regional Archives.*

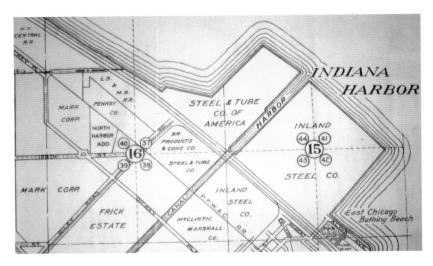

The industrial expansion into the lake, circa 1940. North
Township, *index map. Calumet Regional Archives.*

The new slabbing mill on Inland's eastward expansion into Lake Michigan,
March 1957. *Calumet Regional Archives, Inland Steel Photograph Collection.*

Inland and Youngstown plants showing Inland's 1950s expansion into Lake Michigan. The Plant 2 peninsula (right) grew to more than seven hundred acres (larger than a square mile). Inland is to the right of the canal, Youngstown Sheet and Tube to the left. *Calumet Regional Archives, Inland Steel Photograph Collection.*

Inland's expansion continued through World War II and the prosperous postwar era. The company expanded in the 1950s and again in the '70s, when it added 750 acres in a $2 billion "northern expansion" into the lake.[14] In 1998 Inland Steel was purchased by Ispat International to create Ispat Inland. Six years later Ispat merged with LNM Holdings to form Mittal Steel, and in 2006 Mittal and Arcelor merged, creating ArcelorMittal.

BUFFINGTON HARBOR:
UNIVERSAL ATLAS CEMENT COMPANY

In 1903, development moved further east as the Cement Department of the Illinois Steel Company began construction on the first of several cement plants at Buffington Harbor—just east of Indiana Harbor. This plant used slag, a waste product from the steel mills, and limestone from

Lake Huron to make cement. Due to increased demand for cement, a second mill was built there in 1906.

With Portland cement such a valuable commodity, Illinois Steel separated the company's Cement Department and created the Universal Portland Cement Company, which it continued to own. In 1912 a third mill was erected at Buffington, making the facility, with a daily output of twenty-three thousand barrels, the world's largest cement works.

On June 10, 1927, the company officially opened its new fifty-five-acre Buffington Harbor, which was at that time Lake Michigan's deepest. US vice president Charles G. Dawes was present and raised the flag at the port dedication, which was attended by more than three thousand people.[15] The company merged with the Atlas Cement Company in 1930 and became the Universal Atlas Cement Company. It stretched for several thousand feet along the Lake Michigan shoreline and could produce 10,250,000 barrels of Portland cement annually. The plant had about five hundred employees. Although it was located at the lakefront, when the wind was from the northeast, residents of Indiana Harbor were reminded time and again how close the plant was to their neighborhoods, as yards, cars, and even interior furniture could be lightly dusted with gray cement dust.

GARY AND U.S. STEEL

When U.S. Steel representatives visited the site of their proposed mill south of Lake Michigan in 1905, they realized that much work would have to be done to reshape the land to make it compatible with their plans. Dune-and-swale topography, with its alternating, long, and parallel sand ridges separated by wet swales, made even moving equipment difficult. Making the land conform to the company's plans required massive manipulation.

Work began in 1906. The sand ridges were scraped. The wetlands were drained and filled. Rail lines were moved. A harbor with a mile-long canal and turning basin was dredged. All in all, in order to build all the components of this huge complex, eleven million cubic yards of sand were excavated. Rumors flew that more material was moved to build U.S. Steel's mill than was moved to dig the Panama Canal, but even

The Universal Portland plant before its 1930 merger with the Atlas Cement Company. The postcard gives the location as Indiana Harbor, Indiana, but in reality the plant was in Gary, at its far west end, close to Indiana Harbor. *Steven R. Shook.*

Gary's mayor-to-be Thomas Knotts, Hammond's former mayor and U.S. Steel's attorney Armanis (A. F.) Knotts, and Mr. Gastel, probably a U.S. Steel official, looking for a site for Gary's first well at what is now Fifth and Broadway, April 18, 1906. *Calumet Regional Archives.*

Excavation for U.S. Steel open hearths, August 1906. *Calumet Regional Archives.*

though eleven million cubic yards was not even close to the 269 million tons moved in building the canal, it was still a huge amount of material.

Although the company owned nine thousand acres of land east of Buffington, the configuration was not exactly what was wanted. A major problem was the location of the Grand Calumet River, which flowed through the southern portion of the plant site. So it was decided to move the river one quarter mile to the south. Then when more land was needed on the north side, the shoreline was extended seven hundred feet northward into Lake Michigan.

Fifty-six open-hearth ovens, eight blast furnaces, a coke plant constructed on the east side, and various mills for making bars, plates, and rails were all built. The American Bridge Company plant and the Gary Sheet and Tin Mill, with its modern Galvanizing Department that finished many of the products with a thin coat of zinc, were built west of Gary Works in 1909 and 1910, respectively. Both were subsidiaries of U.S. Steel.

Opening the new channel for the Grand Calumet River,
May 2, 1907. *Calumet Regional Archives.*

While all of this was going on, the company created the Gary Land
Company to make plans to develop the city of Gary. Streets were laid
out in a grid pattern south of the new channel for the Grand Calumet
River, with the main north–south street, Broadway, leading directly to
the plant gate. Sewers and water mains were built. Streets were paved.
Lots were sold. Boarding houses were built. And in a rather short time
an industrial city was sitting where two years earlier there had been sand
ridges and wet swales. Just two years later, the first ore boat entered the
Gary harbor, greeted by hundreds of plant employees and business and
government leaders.

By the 1920s, Gary Works, the country's largest steel mill, had installed four more blast furnaces and had more than sixteen thousand

employees. The "Steel City" of Gary grew along with the company and by 1930 it had a population of more than one hundred thousand.

PORTER COUNTY

Sand mining has had a huge impact on Porter County. This county, along with areas of eastern Lake County and western LaPorte County, was where the sand was most plentiful. Sand mining in Porter County, with the vision of a future port and heavy industry, was started about 1880. For decades, the Knickerbocker Ice Company, the Consumers Corporation, and dozens of other similar but smaller corporations were removing sand from areas adjacent to the lake. Because of sand mining alone, the Dune Park rail station of the Michigan Southern Railroad (at what is now the Port of Indiana) was the line's most profitable freight station between Elkhart and Chicago. Hundreds of carloads of sand were hauled away each year.

For more than fifty years, government officials and business owners were envious of the heavy industry in Lake and LaPorte Counties. They certainly would have liked some of the tax money paid by those industries to help finance projects in their own counties.

In 1920, Inland Steel purchased nearly a square mile from the Ogden estate along the lakefront, just east of the Lake/Porter county line. Inland officials stated that the land was needed for possible expansion, but in fact Inland's only benefit from owning the land was the revenues made from continuing the sand-mining operations begun much earlier. It never did build a Porter County mill, much to the chagrin of the county's movers and shakers, and eventually, in 1967–1968, it sold the land to the National Park Service.

In 1930 the National Steel Company announced that it planned to build a steel plant at Burns Ditch, but the Great Depression forced the company to postpone its development. However, its plans were enough to cause NIPSCO to recognize that the heavy industry expected to be established would require much more electricity than it was able to supply. Thus in 1932 it purchased 350 acres of duneland east of the proposed plant.

In 1935, the Northern Indiana Industrial Development Association was formed to convince others, including members of Congress, of the advantages of a deep-water port. The creation of the Save the Dunes

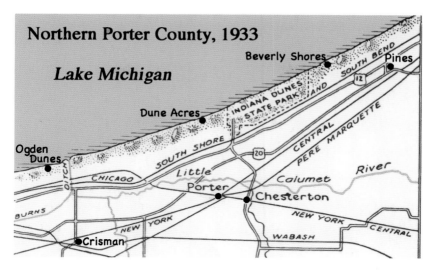

Northern Porter County in 1933. Adapted from Clark Ray.
Moore, Calumet Region, *between 336 and 337.*

Council in 1952 did nothing to decrease the association's enthusiasm. In fact, in the mid-1950s as the United States and Canada jointly planned the opening of the St. Lawrence Seaway, enthusiasm increased. The association promoted the thousands of jobs that the port would support, not to mention the large amounts of tax money that would benefit Porter County. Port opponents lobbied to save the ecologically important Dunes area. Then in 1956, the Bethlehem Steel Company, assuming that the efforts to build a public port would be successful, began purchasing land east of the proposed port.

In the late 1950s Midwest Steel began construction on a finishing plant on the east side of Burns Ditch. Even without a port, it opened its plant in 1959. Midwest owned land on both sides of the ditch but used the west side for settling basins for its pickling liquids and its waste treatment plant.

In 1960 NIPSCO began construction on its second lakefront coal-fired generating station. It named this one after Joseph Bailly, the Northwest Indiana pioneer who once dreamed of building a Lake Michigan port near this site. Two years later, Bethlehem announced that it also intended to build a mill. Assured that Congress would soon approve funding for a port that it could use, Bethlehem arranged for the Missouri

Midwest Steel's plant in 1961. Its sanitation plant and settling
basins (now Portage Lakefront and Riverwalk) are to the left
of Burns Ditch. *US Environmental Protection Agency.*

Valley Dredging Company to remove two and a half million cubic yards
of sand. Construction began soon thereafter.

Congress approved both the Port of Indiana and the Indiana Dunes
National Lakeshore in 1966. Construction on the port began almost im-
mediately. The first ore boat arrived in September 1969, and Bethlehem
produced its first steel in December of that same year.

INDUSTRIAL CONSEQUENCES

The growth of the Calumet Area was an amazing phenomenon. In a
rather short time, Northwest Indiana changed from a vast wet and sandy
wilderness, considered worthless, to an industrial powerhouse. Calumet

Bethlehem Steel's Burns Harbor facility, then the company's largest plant, circa 1970. Today the facility is owned by ArcelorMittal. The Port of Indiana can be seen in the upper left corner. *ArcelorMittal.*

Standard Oil refinery looking south. *Steven R. Shook.*

Gary Works, circa 1955. *Steven R. Shook.*

Area industries gave jobs and prosperity to thousands of people and played a large part in the national efforts during both world wars.

But with no environmental regulations, or at best insufficient ones, sewage from the municipal sanitary districts and contaminants from numerous industrial factories were dumped directly in the Calumet River, while smoke, also laden with dangerous chemicals, was sent up smokestacks. These local practices, however, were not much different from what was being done in other areas all across the country.

Air Pollution

Although nearby residents complained about dust on their cars and furniture and foul-smelling air, the medically harmful effects of many of the pollutants were not yet known. Pollution was the price a community paid for prosperity.

U.S. Steel's coke plant, near the end of the company's property, was the dirtiest part of Gary Works. Coke was essential to the making of

Smoky skies in northern Lake County, circa 1959. *Calumet Regional Archives.*

steel, and several processes were essential in making the coke. Unfortunately, each step emitted pollutants into the air. As coal was heated to nearly two thousand degrees Fahrenheit, carbon particles, hydrocarbons, carbon monoxide, methane, and sulfur dioxide all escaped into the air. The cooling process then resulted in steam containing ammonia and phenol that also escaped into the air. Unwanted liquid waste flowed through sewers into the Grand Calumet River.

The dust in the coke plant was so thick at times that employees had difficulty seeing. The workers, their clothes, and their equipment were all covered with a reddish-brown layer of grime. Employees who worked in the mills often drove old automobiles to work, leaving their newer, better cars at home where they wouldn't get covered in dirt. In his book *Inland Steel at 100: Beginning a Second Century of Progress,* Jack H. Morris explained that dust at the steel mills "came in three colors—red from iron ore, black from coal and coke, and white from limestone used as flux in the blast furnaces [as well as from the cement plants]. Soot billowed

A northeast wind blows smoke into the lakefront cities. *Calumet Regional Archives.*

from the furnaces and ovens while pickling lines and zinc galvanizing pots emitted a foul stench. 'If cleanliness is close to godliness,' one
early observer commented, 'then these men must be one step from
perdition.'"[16]

The cement plant at Buffington sent into the air dust made of fine
particles of slag and limestone which were crushed in the cement-making
process. Residents of the nearby communities in Indiana Harbor con-
stantly complained about the cement dust that coated their automobiles
outside and furniture inside.

The worst industrial pollutants were sulfur dioxide, nitrogen oxides,
carbon monoxide, volatile organic compounds (VOCs), and suspended
particulates. The vast majority of all five of these air pollutants came
from the large steel and petroleum companies and from the nearby lakefront generating station.

Air pollution affected all residents and visitors to the Calumet Area.
The largely uncontrolled pollutants were blamed for respiratory illnesses,
asthma, and emphysema. In 1910, the city of Gary enacted a smoke con-

trol ordinance, but it did not apply to industries north of the Grand Calumet River and was therefore of limited usefulness.

Water Pollution in Lake Michigan

The effects of pollution in Lake Michigan were noticed in LaPorte County as early as the 1860s when the fishing industry there began to suffer. Harmful pollutants started entering Trail Creek and Lake Michigan as early as the 1830s when a sawmill, the Michigan City area's first industry, was built in 1834 alongside Trail Creek.[17] It used water from the creek to run its equipment, and it dumped its waste (including huge amounts of sawdust) into the creek—downstream of its plant, of course. Meanwhile the town was developed and its wastes were also directed, largely aboveground by gravity, into that same stream.

Water quality at the western end of Indiana's lakefront began to suffer beginning in the late nineteenth century and greatly increased in the early twentieth century as Lake County industries and cities grew along the lakeshore. Then in 1925, when the Indiana Harbor Ship Canal was finally connected to the Grand Calumet River, its polluted river waters also began to flow into Lake Michigan.

The lake's waters got so bad that in 1943, the city of Chicago sued Gary, East Chicago, and Hammond, complaining about the "great amount of undisinfected filth, sewage and poisonous and unhealthful and noxious matter" spewed out by Indiana industries. Chicago, of course, produced much more sewage and industrial waste than did Indiana, but the city had caused the Chicago River to reverse its direction of flow and so gladly sent its sewage to downstream Illinois sites. Thus city officials were upset when the waters of Lake Michigan were polluted by Indiana.[18] (In all probability, the offending sewage was more likely to have come from Milwaukee, because the normal lake current tends to move southward along the western shore.)

Smoke and soot affected the lake as well. Bathers at the Lake Michigan beaches became accustomed to seeing black specks floating on the water and littering the otherwise white sandy beaches. Indeed, many of the chemical pollutants in Lake Michigan came from the air.

It should be noted here that in the early days, the three Indiana municipal waterworks that drew water from Lake Michigan merely pumped raw lake water into the city. Chlorine wouldn't be added to city water until later.

In 1936, just eleven years after water from the Grand Calumet River started flowing into the lake through the Indiana Harbor Ship Canal, the Water Department superintendent in Hammond wrote bitterly about the polluted water that the department was pumping out of the lake for its citizens. He claimed that the major portion of that pollution was caused by the discharging of raw sewage from the northern part of his own city of Hammond as well as the city of Whiting, directly into Lake Michigan only a quarter mile from the department's water intake. He added, "Considerable pollution also reaches indirectly through the Little and Grand Calumet rivers and through the Indiana Harbor Ship Canal."[19] Until that year, the Hammond Water Department had not filtered the water, but the pollution in the lake had forced the city to build a modern filtration plant. The superintendent went on to argue that if the cites of Whiting, East Chicago, Gary, and his own Hammond "would build sewage treatment plants, a considerable portion of the human sewage pollution would be removed from Lake Michigan."[20] Gary didn't start filtering its water until the early 1950s when the city council gave a contract to the Gary-Hobart Water Corporation to take over the water supply system for the two cities.

In 1987, a Lake Michigan advisory suggested that catfish, carp, trout longer than twenty-three inches, and Chinook salmon longer than thirty-two inches caught in Lake Michigan's Indiana waters not be eaten. Certain smaller fish should not be eaten by young children, nursing mothers, pregnant women, or women who might become pregnant. And no one should eat more than one meal of Indiana Lake Michigan fish per week.

Different Times, Different Attitudes

In the days before 1970 there was no EPA and the word "sustainability" had not yet been used for anything except forestry management. Only scientists used terms like "particulate matter," and although families enjoying the waters of Lake Michigan on hot sunny afternoons might not have liked flakes of soot floating on the water, they didn't consider the effects of *Escherichia coli*. Smoke coming out of industrial stacks was a sign of prosperity. People were employed. Dirty air and water were certainly not liked, but they were largely tolerated. In fact, because of the widespread practices of pipe, cigar, and cigarette smoking, the air inside some people's houses was often smokier than it was outside. Pipe smoking was sophisticated. A "good" cigar was a treat. In spite of the fact that the surgeon general had warned Americans about the dangers of smoking, many people apparently were either ignorant of or indifferent to its risks.

This isn't to say that no one cared that the air and water were dirty. In 1950 a city councilman from the Glen Park section of Gary raised the issue and asked why the city had no meaningful smoke-control ordinance. Gary mayor Eugene Swartz responded that the issue was very complex and warned the city council against taking any hasty action. The issue was dropped.

Decades earlier, in 1923, Gary superintendent of schools William Wirt developed Dune Acres in Porter County as a community where one could live away from the dirt of the city. In 1941 U.S. Steel opened the Good Fellow Camp in the Dunes so that the children of millworkers could spend a week away from the city's noise and dirty air. It was a given that the air in industrial areas was dirty. However, rather than work to find ways to clean that air, the thinking was that one would just have to leave dirty areas in order to breathe fresh air.

"Scene on Grand Calumet River, Miller, Ind." A postcard from the early twentieth century. *Genealogy Trails History Group.*

Industrialization of the Grand Calumet River and the Indiana Harbor Ship Canal

5

MANY OF THE HEAVY INDUSTRIES THAT WERE LOCATED SOUTH of the lakefront area were in Lake County and bordered either the Grand Calumet River or the Indiana Harbor Ship Canal. By 1960 the Grand Calumet River was nothing like it was in 1860.

At that earlier time, the river was a slow-moving, clear and unpolluted, westward-flowing stream. Near its headwaters, the lagoons in Miller, was the Miller (Calumet) Gun Club, a hangout for wild-game hunters. Near the state line was the Hohman Bridge, one of the only bridges to cross the river. In 1860, it stood next to a six-room log cabin and inn just north of the river run by Ernst and Caroline Hohman, the area's first settlers.

HAMMOND

In the mid-1860s George Hammond, the owner of a meat market and slaughterhouse in Detroit, had a rail car specifically designed to ship meat. It was insulated and had built-in containers for plenty of ice to keep the car's interior cool. He then became the first merchant in the United States to ship meat in a properly refrigerated rail car. Because he owned the patent, he had exclusive rights to all such cars.[1]

So pleased was Hammond with this new type of transportation that in 1869 he decided to open a plant in the Chicago area to take advantage of its stockyards and abundance of potential customers. He and his partner Marcus Towle found a rather vacant spot just east of the Illinois-Indiana state line. The Indiana location was close enough to the Chicago stockyards, adjacent to the Grand Calumet River where every winter his crews would cut and store ice, and far-enough away from any populated areas

George Hammond.
Hammond Public Library.

that would object to the smells that his business gave off.

To keep taxes and freight costs down, Hammond decided to keep the facilities at his State Line Slaughter House looking shabby, as though (if the rates were too high) the plant would move. Besides the cutting houses, Hammond and Towle built a large icehouse to store the ice harvested each winter from the river.

They also built boarding houses for their employees, but many of the men kept their families in Chicago as the stench around the plant was not conducive to family life. Towle lived nearby and in later years became the first mayor of the city of Hammond. George Hammond never lived there.

The firm started with just eighteen employees but it still was able to ship at least three cars of beef to Boston each day. By 1891 it had one

A scale model of the Hammond Company refrigerated rail car. *Steve Coxhead.*

The G. H. Hammond Company

PACKERS

OF

Beef, Pork and Canned Meats

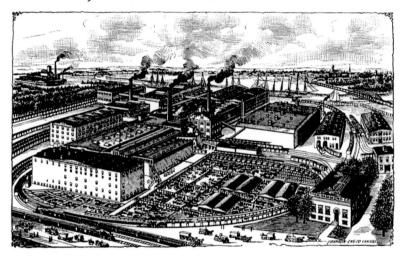

MANUFACTURERS OF

"CALUMET" BUTTERINE

HAMMOND, IND.

A Hammond Company full-page advertisement in the 1897 *Lake County Blue Book*. Butterine was imitation butter made from animal fat. *Calumet Regional Archives.*

thousand employees and was processing three hundred thousand cattle, twenty-five thousand sheep, and ten thousand hogs annually.[2] By the time George Hammond died in 1886, he was the owner of a $3 million corporation and eight hundred patented refrigerated rail cars. The plant remained open and busy until October 23, 1901, when its wooden buildings were destroyed by a huge fire. The plant's remains passed into a trust that decided not to rebuild. Only 12 families lived in the Hammond area when the slaughterhouse was started; 12,376 lived there when it closed.

The Calumet River's first major pollutants came from the State Line Slaughterhouse. The firm produced prodigious amounts of waste—primarily in the form of the carcasses of its slaughtered animals. Many of these carcasses were simply dumped in the river in the hope that they'd be carried away. Some may have been, but others, as vividly described by historian Lance Trusty, "bobbled along the Grand Calumet en route to Lake Michigan, collected along the banks, or just sank to the bottom. Summer was memorable in old State Line."[3] Historian Powell Moore noted that when one was downwind of the plant the odors were almost unbearable.[4]

MORE HEAVY INDUSTRY ARRIVES

Hammond was, and still is, the eastern gateway into Chicago, and by 1875 a dozen rail lines were built across the city—all heading into or around Chicago. The Gibson yards were located in the middle of the city, crossed by the nine-span bridge that was said to be the country's longest bridge over just land.

Land in Indiana was cheaper than in Illinois and thus many other businesses, including both heavy and light industries, moved into Hammond in the years that followed. Marcus Towle's Kingsley Foundry opened in 1892, doing business later as Mackie Steel Tube.

In 1898 the Simplex Railway Appliance Company built its factory near Hohman Avenue on the north bank of the Grand Calumet River. (The packing company was on the south bank.) Its employees fabricated freight car suspensions and brake parts for locomotives. Seven years later, the company merged with the American Steel Foundries, which also had a plant in East Chicago. The company remained in the

Postcard featuring the Simplex Railway Appliance Company
(later the American Steel Foundries) on the north bank of the
Grand Calumet River. *Hammond Public Library.*

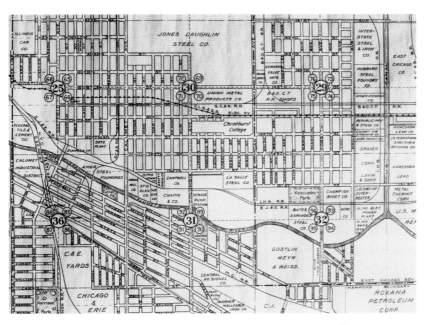

The Grand Calumet River in West Hammond, circa 1940. The city created residential
neighborhoods south of and right next to the Grand Calumet River, but the north
bank was industrial. North Township, *index map. Calumet Regional Archives.*

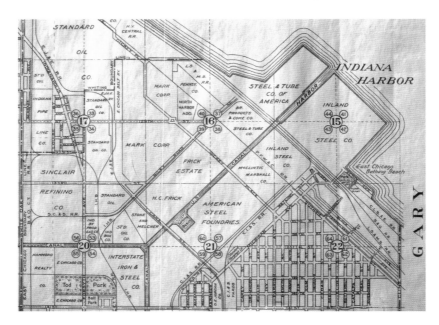

The Indiana Harbor Ship Canal (circa 1940), showing both the west and south forks. The residential community in the lower right corner is called Indiana Harbor. The Mark Manufacturing Company, later Youngstown Sheet and Tube, had purchased the Steel and Tube Company and tripled its capacity by building out into the lake. North Township, *index map. Calumet Regional Archives.*

downtown Hammond area until 1975.[5] The Fitzhugh Luther Company, meanwhile, produced steam shovels and heavy railroad equipment.

The major industrial area along the river in Hammond was near the Hohman Avenue Bridge over the Grand Calumet, the site of the Hammond Meat Packing Plant. In 1874 J. M. Hirsh started a small factory there that morphed into Hirsh, Stein & Company, later the United Chemical and Organic Products Company, a glue and fertilizer factory.[6] According to Powell Moore, by 1915 it employed about four hundred men and was said to produce 5 percent of the glue used in the United States.

THE INDIANA HARBOR SHIP CANAL

In the year 1900, the only Indiana harbor on Lake Michigan was at Michigan City. The planners of the city of East Chicago believed that its industrial development depended upon the construction of a ship

The Indiana Harbor Ship Canal as seen from the top
of the Cline Avenue Bridge, circa 1988.

canal to provide docking facilities for its plants and a belt line railroad
to connect these industries with other rail lines in the Chicago area. The
East Chicago Company, which controlled five thousand acres in the East
Chicago area, started construction on the water facilities. The New York
Central Railroad built the Indiana Harbor Belt Line. Work on the outer
harbor began in 1901.

The canal is two hundred feet wide and twenty-one feet deep. It
extends inland in a southwestward direction for one and a half miles,
at which point two forks extend it farther: a west fork reaches to Lake
George in Hammond and a south fork intersects the Grand Calumet
River. The canal dredging was begun in 1903, but it took six years to
complete the first mile of the project. And it wasn't until 1925 that the
canal finally connected the middle of the Grand Calumet River to Lake
Michigan. The canal and both forks were originally intended to facilitate
transportation by ship; however, the industries that located along the far
south fork had no need of docking facilities, and so that portion of the
canal has not been maintained for ship travel.[7]

THE GRAND CALUMET RIVER IN EAST CHICAGO

Between Cline and White Oak Avenues, the Grand Calumet River forms the boundary between Hammond to the south and East Chicago to the north. This part of the river became heavily industrialized during the first half of the twentieth century. Because the area contained the Grasselli Chemical plant and a large number of petroleum companies, some called it "Chemical Alley."

As soon as the Ship Canal connected the river with Lake Michigan, the current's direction in the part of the river west of the canal changed from flowing west into Illinois to flowing east into the canal. From there, the waters from both directions then flowed north directly into Lake Michigan.

POLLUTANTS IN THE RIVER AND CANAL

In the years from 1869 to the middle of the twentieth century a large number of industrial companies built factories next to the Grand Calumet River, and most if not all of them discharged their wastewater directly into that river. This practice was not unusual; in fact it was practically universal. The river was thought of as an extension of the sewers, and little thought was given to what happened to the unwanted chemicals after they left the plant. (Except for the volume of the materials discarded, this wasn't much different from private citizens dumping unwanted medicines, antifreeze, or other possibly harmful chemicals down the drains in their homes.)

For centuries, when cities modernized and built sewers, those sewers dumped their contents into a nearby river—downstream of town, of course. So, naturally, the cities of Hammond, East Chicago, and Gary all dumped their storm and sanitary sewer effluent directly into the river. In those days, when "dilution was the solution to pollution," this practice was neither illegal nor considered morally wrong. There were few enforced regulations against polluting rivers. However, all these contaminants had great negative impacts on the slow-moving river as well as the grounds adjacent to it.[8] In addition, each week the mills, foundries, refineries, and other plants poured millions of gallons of sometimes oily wastewater into the Grand Calumet River and the Indiana Harbor Ship

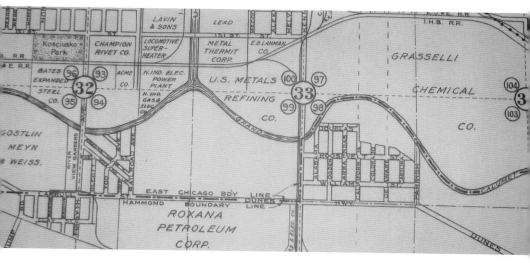

The Grand Calumet River in East Chicago and West Hammond, circa 1940. North Township, *index map. Calumet Regional Archives.*

East Chicago's Grasselli (misspelled here as "Gressalli") Chemical Company plant on the Grand Calumet River. The Grasselli plant and property were later purchased by DuPont. *Steven R. Shook.*

Oil-soaked vegetation in the Grand Calumet River, circa 1983. *Dorreen Carey.*

Canal. Included in that amount were thousands of pounds of ammonia, nitrogen, phenol, and cyanide, and small particles of iron, manganese, and chromium.

The oils floated on the river water while the heavy solids sank and remained on the riverbed. Nevertheless much of the waste followed the course of the river and canal and ended up in Lake Michigan, the source of drinking water for all the lakeside communities on both sides of the state line. Although the pollutants were, of course, diluted once they reached the lake, as early as 1933 the southern portion of Lake Michigan was described as "having one of the nation's most serious water pollution problems."[9]

The Grand Calumet River is a river with no tributaries. By the middle of the twentieth century 90 percent of its water came from municipal and industrial wastewater outlets. The Grand Calumet River essentially had become a dead industrial sewer. No fish could survive its waters. Even after the cities built expensive sewage treatment plants, periods of heavy rainfall would overload the system and raw sewage would flow into the river. And as the river eventually flowed into Lake Michigan, the lake's

Reporter Casey Bukro's oil-drenched hand at the Indiana Harbor Ship Canal.
Photo by Luigi Mendicino, Chicago Tribune/TNS.

health suffered, and so naturally did the lake-centered commercial fishing industry.

In 1967, to prove his point that the water was badly polluted, *Chicago Tribune* reporter Casey Bukro dipped his hand into the oil-coated Indiana Harbor. He wrote, "A stream of poison pours into Lake Michigan from the Indiana harbor—a contender as champion of polluted waterways. . . . Indiana Harbor receives staggering doses of industrial pollution from the Indiana Harbor Ship Canal and the Grand Calumet River. . . . These are the sewers for industry here. The waters cry for help."[10]

Not all of the pollution in the river has entered it through "point sources," that is, through industrial and municipal wastewater discharge pipes or sewers. Much of it is from "nonpoint sources" such as industrial waste site runoff, salt and tire residue from streets and highways, residential lawn fertilizers and weed killers, and leakage from underground storage tanks, hazardous waste sites, and potential Superfund sites. It is estimated that five to ten million cubic yards of contaminated sediment has come in via these nonpoint sources.[11] Much of it either washed in during rainstorms or seeped in through the sandy soils. By the 1970s, because of contamination from all of these various sources, the Grand Calumet River and Indiana Harbor Ship Canal were deemed to be among the most polluted rivers in the Lake Michigan basin.

THE PUSH FOR PARKS AND DUNELAND DEVELOPMENT, 1890–1929

6

ONE OF THE EARLIEST GROUPS TO PROMOTE OUTDOOR RECRE-
ation was the Playground Association of America. President Theodore
Roosevelt was the honorary chairman of this group and Chicago social
worker Jane Addams was its vice president. Soon after the association
held its second annual meeting in Chicago, a number of Chicago-area
members who enjoyed the outdoors joined forces with several other local groups including the City Club, the Geographic Society of Chicago,
the Audubon Society, the Art Institute Instructors, and the Women's
Outdoor Art League, and organized "Saturday afternoon walking trips."

Recreational hiking was a new idea at the time, and when the organizers first used the word "hike," they put it in quotation marks! And unlike
many other health-oriented activities, its hikes were coed; both young
men and women participated in and led them. The first organized walk,
on April 18, 1908, was a three-mile trek near Willow Springs and was attended by 128 hikers.[1] Five hikes were conducted that first year.

The group's May 1911 circular used the name "Prairie Club" for the
first time. The name is said to have been proposed by Jens Jensen, a
widely known landscape architect and early president of the group.[2] In
early July 1908, with the South Shore Railroad line completed and a station at Tremont opened, a trip from Chicago to the Indiana Dunes was
greatly simplified, and soon the Dunes became a favorite destination of
the club.

In 1916, the National Park Service was established and Prairie Club
member Stephen Mather was named its first director. At that time, all
of the national parks were west of the Mississippi River and, not sur-

Hiking in the Dunes, May 11, 1915. Note the ankle-length dresses worn
by the women. *Arthur Anderson/Calumet Regional Archives.*

prisingly, Prairie Club members decided that the Dunes would be an
appropriate addition to the new national park system. Thus that same
year, with the goal of establishing a national park in the Dunes, Prairie
Club members created the National Dunes Park Association (NDPA).

The Prairie Club's Conservation Committee, headed by Catherine
Mitchell, published articles that described ecological scientific findings
as well as the attention the national park campaign was receiving. The
best publicity the committee was able to arrange was an article about the
Dunes in the May 1919 issue of *National Geographic* magazine, which featured (naturally but unfortunately) black-and-white photographs sub-
mitted by club members. In 1917, the committee printed and distributed
more than one hundred thousand decorative "Save the Dunes" stickers.

The Prairie Club's beach house on Lake Michigan. *Calumet Regional Archives.*

Duna, the Spirit of the Dunes, a 1913 "masque" to celebrate the opening of the beach house. *Prairie Club of Chicago.*

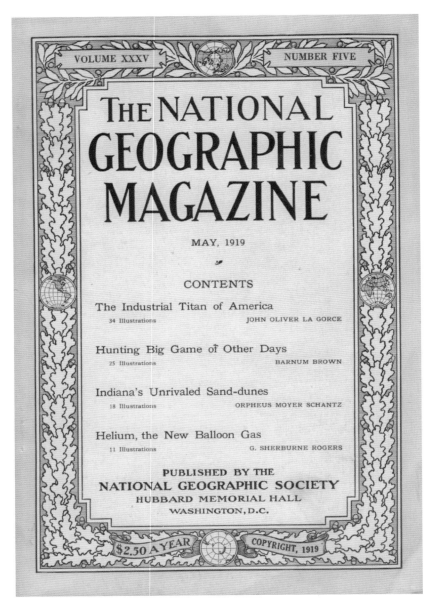

VOLUME XXXV   NUMBER FIVE

The NATIONAL GEOGRAPHIC MAGAZINE

MAY, 1919

CONTENTS

PUBLISHED BY THE
NATIONAL GEOGRAPHIC SOCIETY
HUBBARD MEMORIAL HALL
WASHINGTON, D.C.

$2.50 A YEAR   COPYRIGHT, 1919

Cover of the May 1919 issue of *National Geographic*.

"Save the Dunes" stamps of 1917. *Save the Dunes.*

Club member Henry Chandler Cowles presented a lecture that included the findings of his biodiversity research in the Dunes, and Mitchell assembled an exhibit that was displayed at Marshall Field's department store. Later, the exhibit "went on tour" and was displayed at several Women's Clubs conferences throughout the Midwest.[3]

In the days before television and radio, pageants were a popular form of entertainment. So in February 1917, when NDPA members formed a Dunes Pageant Association, they commissioned pageant writer Thomas Wood Stevens, head of the Drama Department at the Carnegie Institute of Technology, to compose a grand one for the Dunes. Called *The Dunes under Four Flags*, the extravaganza would be held on both May 30 (Memorial Day) and June 3. It would be held outdoors right on the Lake Michigan shoreline and would weave poetry, music, and dance around the themes of Duneland history and preservation.[4]

The "Jensen," or "Big Blowout," at Waverly Beach (today west of the pavilion) seemed to be the best place to hold the pageant. Directors, choreographers, seamstresses, speakers, choirs, and bands were all enlisted and actors found who could play the parts of American Indians, Marquette, fur trappers, British and Spanish soldiers, pioneers, and even Daniel Webster.[5]

NDPA secretary and Gary resident Bess Sheehan happened to be the president of the Indiana Federation of Women's Clubs and she arranged for various Women's Clubs to make and distribute newsreels for use in movie palaces across the country. Newspapers also provided great publicity. In addition to the Prairie Club, event sponsors included approximately thirty Women's Clubs from around the Midwest.[6]

The situation and the national mood changed on April 6, 1917, when the United States entered World War I. Very quickly national priorities changed from domestic issues to winning the war. But the NDPA knew that a nation at war cannot and should not forget homeland issues, and it wanted to make sure that the local population remembered that the Dunes still needed protection.[7] Therefore the association decided not to cancel the pageant, but instead changed the script to give it a patriotic flavor. Parking was arranged for five thousand cars, and extra trains on the South Shore Line were arranged.

View from "backstage" at the Sunday afternoon Dunes
Pageant, June 3, 1917. *Prairie Club of Chicago.*

Marquette Park Beach, July 1941. *K. M. Bain family.*

Unfortunately, the Memorial Day performance was rained out, but the Sunday extravaganza was seen in all its splendor by about twenty-five thousand people. In spite of the apparent success of the pageant, as the war continued it became obvious that Congress was in no mood to deal with creating a new national park. Besides the war, another major concern was the fact that up to this time, all national parks had been created out of land that the US government already owned, and the Dunes were all privately owned. Thus Congress would have to appropriate funds for purchasing the Dunes, over the objections of several Porter County business folk who opposed the park. Meanwhile sand mining of the Dunes continued. By the time the war ended in 1919, the enthusiasm was gone, and on May 15, 1919, the NDPA admitted defeat and suggested that Indiana residents try to have a state park created instead.

Two NDPA officers, however, continued their work. William P. Gleason, president of the organization, and also president of the Gary Park Board and superintendent of U.S. Steel's Gary Works, worked to arrange the creation of Lake Front Park (now Marquette Park) in Miller. And Bess Sheehan worked tirelessly for several years and eventually, and with the help of other Women's Clubs across the state, convinced the Indiana General Assembly to create Indiana Dunes State Park. The two parks together were far too small to hold the crowds of people who wanted to go to the beach on summer weekends, but they were a start.

Today, the Prairie Club that started all this still exists, as described by Ryan Chew in *Chicago Wilderness* magazine, as a "social group with a social conscience."[8] The club still honors its goals of conservation, education, and enjoyment of nature. Its some eight hundred members still support environmental efforts through its Conservation and Education Fund. A percentage of its membership dues goes to support parks, prairies, and the Dunes.[9]

THE DUNES HIGHWAY AND DUNELAND DEVELOPMENT

In 1918, before World War I was over, a group of Northwest Indiana businessmen organized the Dunes Highway Association and started lobbying for the construction of a new direct highway between Gary and Michigan City. It took a while, but the lobbying was successful and construction began in March of 1922.[10]

The Dunes Highway (probably in eastern Porter County), circa 1923.
Calumet Regional Archives.

The eastern portion of the highway was easier to build than its western section. That section, from Michigan City to Baillytown, basically used the same route as the old Chicago–Detroit stage road, which eighty-five years earlier had been laid out along the sand ridge of Lake Michigan's ancient Calumet Shoreline. The western section, from Baillytown to Miller, had to cross the soggy Great Marsh. The highway was completed and opened to the public in November of 1923. Christened the "Dunes Highway," it was northern Porter County's only concrete highway. With no ground-level railroad crossings, and connected with Michigan highways leading to Detroit, this roadway soon became Indiana's busiest.

Several years later US 20 was built south of, and nearly parallel to, the Dunes Highway. Because, as was expected, it diverted much of the traffic from the Dunes Highway, it was nicknamed the Dunes Relief Highway. A generation later, Interstate 94, which also goes from Chicago to Detroit, relieved it still further, so much so that driving the Dunes Highway today can (except at shift change) be a tranquil experience.[11] Many drive it in order to admire the wooded landscape on the National Lakeshore.

Even before the highway opened, speculators foresaw development. Plans were made for residential, commercial, and industrial development in the Dunes. Ogden Dunes and Dune Acres were both platted in 1923, before the highway was even finished. Both communities had plans for a harbor for small pleasure craft, a hotel or guest house, and a golf course

The ski slide at Ogden Dunes, circa 1929. *Mark Stanek.*

with a clubhouse. Both communities also wanted to control their own development, and so they incorporated as towns. Their populations in 2010 were 1,110 and 182, respectively.

For a few years, Ogden Dunes could claim to have the highest man-made ski slide in the country. Just four years after the opening of the Dunes Highway, the Grand Beach Ski Club purchased acreage around one of the high sand dunes north of the highway and constructed a ski jump twenty-two stories high. The club changed its name to the Ogden Dunes Ski Club and hosted an international ski jump competition in 1928 that attracted more than nine thousand people.[12] The next year the South Shore Railroad featured the ski jump on one of its advertising posters. However, attendance dropped once the Great Depression hit, and the attraction was closed in the early '30s.

When Chicagoan Frederick Bartlett platted Beverly Shores in 1927, it was to be a resort community with both commercial and residential areas, larger than any other Chicago-area development along the Lake Michigan shoreline. Unfortunately just three months into lot sales, the stock market crashed. Investors who could afford it purchased some lots but there was little construction. In 1933, Bartlett's brother Robert purchased the development and resumed building. His properties included an impressive hotel, a theater, and a beautiful restaurant on the beach. In 1934 sixteen houses from Chicago's Century of Progress World's Fair were brought to Beverly Shores. The five placed on Lake Front Drive still stand. Of the others, the only one remaining is the home that was originally built to resemble Boston's Old North Church. Beverly Shores was incorporated as a town in 1946.[13]

THE INLAND TOWNS OF THE DUNES

Furnessville predates the Dunes Highway. This unincorporated community alongside Furnessville Road and US Route 20 got its start in 1851 when the Michigan Central Railroad built its line through the area and hired Edwin Furness to manage the local depot. In 1861 Furness became the first "Furnessville" postmaster. In 1929, when the Dunes Relief Highway (US 20) was built through the area, several nineteenth-century farmhouses had to be removed. In spite of the new highway, the area remained a quiet rural community. Not surprisingly, the area attracted many artists, including students from the School of the Art Institute of Chicago. Pulitzer Prize–winning author Edwin Way Teale grew up in Furnessville, and his 1943 book, *Dune Boy: The Early Years of a Naturalist*, described his summers living at his grandparents' Furnessville farm. Today much of the Furnessville acreage north of Route 20 has been incorporated into the National Lakeshore.[14]

Pines is one of Indiana's smallest incorporated towns. Situated along Route 12 atop the sand ridge of the Calumet Shoreline, it was platted as that highway was being planned in 1922–23. However, it was not incorporated as a town until 1952. Two of the oldest motels in the Duneland area are both in Pines, as is Indiana's shortest highway—Indiana State Road 520, just one quarter mile long. In 2010 Pines's population was 217.

Tremont, just south of the state park, was one of the first settlements established in northern Porter County as it was the location of the Green

Tavern, a stagecoach inn, built around 1837 along the old Chicago–Detroit road. The name Tremont refers to Mount Tom, Mount Holden, and Mount Jackson (the last one originally called Mount Green), the three highest dunes along the lakeshore. When the South Shore Railroad was built in 1908, its station at Tremont became for many years the entryway into the Dunes. It also became one of the South Shore's more important stops after the Indiana Dunes State Park was opened in 1926. The station was eventually closed and has been replaced by the new Dune Park Station, which also has the general offices of the Northern Indiana Commuter Transportation District. Tremont was never incorporated as a town. Much of the area today is included within the boundaries of the National Lakeshore.

BURNS DITCH

Burns Ditch was dredged in the mid-1920s in order to lessen the annual flooding of the Little Calumet River. The "Little Cal" has such a low gradient that its waters flow very slowly, much more so than the waters that enter it through Deep River or Willow, Salt, or Sand Creeks. Thus, particularly after a deluge, water enters the river faster than it moves out, often resulting in flooding that inundates farm fields and residential neighborhoods. According to historian Powell Moore, real estate developer Randall Burns spent fifteen years lobbying officials to urge the state to dig a ditch through the tall Tolleston-era sand ridge near where the river was the closest to Lake Michigan. Moore, reflecting the then-current opinion of wetlands, referred to the floodplain as "worthless, except for hunting and fishing."[15] Draining wetlands in those days was referred to as "reclaiming" land.

Civic leaders in the city of Gary supported Burns, but railroad companies objected to the project because they would then have to build and then maintain seven bridges over the water. But Burns was successful in his efforts. Not only was the Little Calumet connected directly to Lake Michigan, but about seven miles of the river itself was straightened. When completed in 1926 the ditch, now called the Burns Waterway, was named for him. Digging Burns Ditch accomplished what it was intended to: the Little Calumet River's flooding was lessened and formerly

Burns Ditch, with the Gary Boat Club in the far right
foreground, circa 1950. *National Park Service.*

inundated lands became available for farming and subdividing. But it
did not eliminate the threat of floods, as the disastrous flood of 2008
demonstrated.

With the construction of the ditch, river water from LaPorte and eastern Porter Counties no longer flows westward into Lake County, but
instead flows north through Burns Ditch into Lake Michigan. Also, river
water in western Porter County and eastern Lake County now flows east
and then also north into the lake. Thus the Little Calumet River now
has the distinction of flowing in two directions. Only the far western
part of the river (from about the Munster/Highland border) still flows
westward into Illinois.[16]

The Burns Waterway allowed the development of many marinas for
pleasure boats near its intersection with the Little Calumet River. The
photo above shows Gary Boat Club boats in the waterway and the club's

The Portage Marina.

building and parking lot to the west (right) on land that later became Midwest Steel's settling basins and is now the Portage Lakefront and Riverwalk.

At the time the ditch was dug, people simply did not realize that wetlands have value. As expected, the creation of this waterway between the Little Calumet River and Lake Michigan drained much of the wetlands south of the Dunes. Perhaps one unexpected consequence of the creation of the Burns Waterway was that agricultural runoff as well as storm water and sewer overflow effluents from the communities along the river in Porter County and most of Lake County now flow directly into Lake Michigan.

PORT VERSUS PARK

Conflict in the '50s and '60s

7

PLANS TO ESTABLISH A PUBLIC DEEPWATER PORT ON LAKE Michigan predate Indiana's statehood. It was this early Hoosier dream that resulted in the moving of the old state boundary line (today at Porter's Beam Street and both Gary's and Chesterton's Indian Boundary Road) from the southern tip of Lake Michigan to a line ten miles north. Fur trader Joseph Bailly planned for a Porter County port in the early 1830s but died before any progress was made. The dredging of Burns Ditch in 1926 for flood relief sparked new efforts to create a public port, but those efforts stalled as well once the Great Depression took hold. In 1929, Midwest Steel purchased 750 acres straddling Burns Ditch. Midwest executives and many Porter County businessmen were hoping that the federal government would build a deepwater port at the ditch. But the Army Corps of Engineers in both 1931 and 1935 rejected that idea because it would benefit just that one company.

So in 1935 a group of businessmen, led by Valparaiso Chamber of Commerce manager George Nelson, planned for a public port east of the Midwest properties and urged Congress to provide funding. Much of that land was owned by the Consumers Company of Chicago, a corporation that was already lobbying Congress for a port that would support heavy industry on the lakefront. Meanwhile NIPSCO, which owned three hundred acres of land just west of Dune Acres, was waiting for the industrial development that would need the electricity that NIPSCO could provide.

Adding to the pressure for industrial development in the late 1950s was the impending opening of the Saint Lawrence Seaway, which would allow oceangoing vessels to come directly to ports on Lake Michigan. Area farmers could imagine their harvests being shipped overseas to customers half a world away.

Meanwhile, a growing number of Northwest Indiana residents were becoming alarmed at the prospect of more of the Dunes being leveled. They had heard that some developers were referring to the Lake Michigan shoreline as an industrial crescent and that some even talked about having the state sell Dunes State Park. Even if that were never to happen, folks could see open space disappearing.

SAVE THE DUNES COUNCIL

So it happened that on June 20, 1952, twenty-one women gathered at the Ogden Dunes home of Dorothy Buell to listen to Bess Sheehan tell the story of her involvement and success in creating the state park thirty years earlier. The women then organized the Save the Dunes Council with the goal of reigniting the national park movement and preserving the natural lakeshore and dunelands.

Dorothy Buell was an amazing woman. As a child, she had visited the Dunes and even participated in the great Duneland pageant of 1917. In 1952 she was sixty-five years old, an age when many people think of slowing down, but Buell had a dream. Speaking for the new organization, she announced, "We are prepared to spend the rest of our lives, if necessary, to save the Dunes."[1] She had what Herbert Read called "a commanding presence, combined with steely determination, dignity, enthusiasm, optimism, and the ability to attract people with the specific skills necessary to accomplish a task."[2]

Sometimes, here in the twenty-first century, it's hard to imagine what these determined folks were up against. The environmental movement hadn't yet arrived. People didn't celebrate Earth Day. Wetlands were considered bad and unhealthy. And the federal government had not yet ever purchased land for a national park. (All the parks were federal properties that simply had never been sold.) Indiana governmental officials and US congressmen and senators strongly favored industrial development.

Yet the need for more recreational lands was unmistakable. In the summer, Gary's Marquette Park and Michigan City's Washington Park were jam-packed. Dunes State Park was so overcrowded that on holidays

Save the Dunes Council members plan their next moves. Standing, from left to right: Merrill Ormes, Walter Necker, Robert Mann, and George Anderson. Seated, from left to right: Sylvia Troy, president Dorothy Buell, and Ann Sims. 1962. *Save the Dunes/Calumet Regional Archives.*

and summer weekends the state police had to turn away cars at the park entrance. Johnson's Beach in Porter sometimes had more than a thousand cars on hot July weekends.

Dorothy Buell started the effort to win over public opinion with dozens of speaking engagements. Photographer John Nelson produced a film describing the Dunes, which was shown to local clubs and service organizations. Members of many groups responded to the call for help. These efforts resulted in more than 250,000 people signing a petition to preserve the dune landscape.

In those early days, it was hoped that the council could raise enough money to purchase the Dunes, and indeed, when the area known as Cowles Bog was put on the market, the council did raise the money and purchased that property. But Duneland was far too large for a small organization to be able to purchase the entire area with donated funds.

DREAMS DELAYED

Believing that a port would soon be built, the Bethlehem Steel Company in 1955 purchased 1,200 acres of open duneland between Midwest Steel's property and Dune Acres. Since US representative Charles Halleck, from DeMotte, in whose district the port or park would be located, and both of Indiana's US senators supported the port, Dorothy Buell contacted Illinois senator Paul Douglas of Chicago. In his younger days, Senator Douglas had spent many enjoyable hours at the Dunes and he agreed to help. On May 26, 1958, as promised, he introduced a bill to establish a national park in the Dunes. At the Senate hearings in Washington, with sixty Save the Dunes Council members in the room and with the walls holding paintings of the Dunes by Frank Dudley (which the council members had brought with them on their bus), council members were informed that bulldozers had begun to level the land that they were trying to preserve. Sylvia Troy, who later served as president of the council, recalled this as the low point of the struggle and noted that the council seriously considered giving up.

Fearing that delay would make building a plant impossible, Midwest Steel began construction on its mill in 1959—even though there was no decision yet by Congress to build a port. The next year, NIPSCO began construction on its coal-fired Bailly Generating Station just west of Dune Acres. However, an Army Corps of Engineers feasibility study recommending a port was found to have errors, and the US Bureau of the Budget decided against endorsing the port and thus halted congressional approval.

Council members visited with all 535 members of Congress, and slowly the council's efforts also began to convince formerly skeptical citizens. Preservation was soon supported not only by environmental groups, but also by area chambers of commerce, the League of Women Voters, unions, and service clubs. Representative Ray Madden told the council that in his thirty years in Congress, he had never before been lobbied by twenty-two groups on one single issue.

Save the Dunes Council member Judson Harris drives Senator Paul Douglas
to a 1961 rally attended by both supporters and opponents of saving the
Dunes. In the backseat are Chicago's mayor Richard J. Daley and Gary's
mayor George Chacharis. *Save the Dunes/Calumet Regional Archives.*

In 1961 Dorothy Buell and Senator Douglas organized a highly pub-
licized tour of the Dunes for Secretary of the Interior Stewart Udall,
National Park Service Director Conrad Wirth, and several area mayors.
Indiana's rather new senator Vance Hartke and soon-to-be senator Birch
Bayh came on board as well.

But by 1962 both dreams seemed thwarted as neither the port nor the
park proposal could get support from a majority of Congress. President
John F. Kennedy, who supported the park, proposed a compromise in
which both the port and the park would be established. His assassina-
tion in 1963, however, doomed the prospects of an early resolution to the
problem. Representative Charles Halleck was the US House minority
leader and was still a strong supporter of the port.

Stewart Udall, secretary of the interior; Senator Alan Bible, chair of the Senate
Interior Subcommittee; Senator Paul Douglas; and National Park Service
Director Conrad Wirth walk through dunes on land owned by the Bethlehem
Steel Company, July 1961. *Save the Dunes/Calumet Regional Archives.*

FINALLY, AN AGREEMENT

In 1966, an agreement was reached in Congress whereby industry, the
port, and the Indiana Dunes National Lakeshore would share Indiana's shoreline. A few months later the Port of Indiana held its official
groundbreaking.

By this time Midwest Steel had already begun operations. Bethlehem
Steel opened in 1967. NIPSCO's Bailly Generating Station followed, and
Indiana's International Port, now the Port of Indiana at Burns Harbor,
was opened in 1969. George Nelson was honored by having the main
street within the port area named for him. The Indiana Dunes National
Lakeshore was to protect eight miles of shore and 8,330 acres of parkland,

including the state park, which could be acquired only if the state would donate it.

With the park authorized, the Save the Dunes Council considered disbanding. However, it was soon discovered that Congress had not authorized any money to purchase land, and so in 1967 another lobbying effort was required to secure acquisition funds. The council's members soon realized that ongoing diligence was needed, and so the Save the Dunes Council, now known simply as Save the Dunes, still works toward the care and preservation of the Dunes.

The duty imposed by sustainability is to bequeath to posterity not any particular thing . . . but rather to endow them with whatever it takes to achieve a standard of living at least as good as our own and to look after their next generation similarly.

ROBERT SOLOW, "AN ALMOST PRACTICAL STEP TOWARD SUSTAINABILITY"

Part Two

Returning to Sustainability

$\mathcal{L}$ike other American communities, Indiana's communities commemorated the first Earth Day in 1970—and have since then made it an annual event. And, like other Americans, Indiana residents wanted their air and water to be clean and healthy. It is probably truthful as well to say that many hoped it would just happen, and happen quickly, without much in the way of sacrifice.

For more than a decade after 1970 Indiana businesses and local governments applied for the required permits to dump waste into rivers and emit it into the air. Various state and local agencies, including county health departments, along with hundreds of businesses, worked with the new Environmental Protection Agency to help get Indiana in line with national priorities. There was still a great concern that environmental controls would be too expensive. Progress was slow.

As late as 1991, environmental problems in Northwest Indiana were still described as "extensive in nature."[1] The area produced more steel than any similarly sized region in the United States. It had four major oil refineries, supplied by six oil pipelines, and eighteen refined oil companies. Area industries, businesses, and residents were emitting tons of pollutants into the river and into the air. More than 180 million pounds of sediment per year entered Lake Michigan at Indiana Harbor alone. Municipal sewers dumped eleven billion gallons of untreated sewage into the Grand Calumet River each year![2] In addition to the heavily polluted Grand Calumet River and Indiana Harbor Ship Canal, the ground itself was polluted. Just the area around the Grand Calumet River and the Indiana Harbor Ship Canal had five Superfund sites,[3] more than four hundred other sites that were requiring cleanup operations, and 462 underground storage tanks, including 150 that were reported to have leaks. The groundwater was undrinkable, as it was polluted as well.

The phrase "sustainable development" was coined in 1987 by the United Nations World Commission on Environment and Development (also known as the Brundtland Commission), which defined the

term as "development that meets the needs of the present without compromising the ability of future generations to meet their own needs."[4] The phrase helped focus the environmental movement, and sustainable development became a worldwide goal. In Northwest Indiana, it helped promote cooperation for restoration of natural areas, cleanup and prevention of pollution, and environmental education.

Very importantly, by this time the bitter confrontations between industry and conservation had become a thing of the past. The two former "sides" had begun to cooperate on a number of ventures. After all, all residents wanted a healthy environment in which to live and raise their families. Clean air and water and sustainability became common goals.

The outlook is becoming quite positive. With expanding cooperation between corporations, environmental advocates, and various levels of government, the possibility of sustainability in this region is certainly increasing.

INDIANA DEPARTMENT OF ENVIRONMENTAL MANAGEMENT

In 1986 the Indiana General Assembly decided that it was necessary to have one state agency to oversee these various efforts, and so it created the Indiana Department of Environmental Management (IDEM). IDEM's mission is to implement the various federal and state environmental regulations while still "allowing the environmentally sound operations of industrial, agricultural, commercial and government activities vital to a prosperous economy."[5] Various departments within IDEM are responsible for air, land, and water quality, and for pollution prevention and technical assistance. IDEM has a Compliance and Technical Assistance Program that offers free and confidential assistance to Indiana businesses and local governments.

IDEM does not work alone, but rather with the EPA and other federal departments as well as citizens, businesses, the Indiana Department of Natural Resources, the Indiana State Department of Health, and other state agencies. Through these efforts much progress has been made in the quality of Indiana's land, air, and water.

Tom Easterly. *Indiana Department of Environmental Management.*

When this book went to press, the IDEM commissioner for the last ten years had been Tom Easterly, who was a resident of Porter County when he was appointed to this position in 2005 by then-governor Mitch Daniels. Commissioner Easterly had been the manager of environmental planning and improvement for NiSource and had earlier been superintendent of environmental services at Bethlehem Steel's Burns Harbor plant. While at Bethlehem Steel, he facilitated it becoming the first steel company to join the Coalition for Environmentally Responsible Economies. Easterly was a founding board member of the Indiana Dunes Environmental Learning Center and served as the second president of its board from 2000 to 2002.

Earthrise, taken December 24, 1968, by the astronauts on the Apollo 8 mission, the first manned voyage to orbit the moon. *NASA*.

Earth Consciousness in the '60s and '70s

8

Never doubt that a small group of thoughtful, committed citizens can change the world; indeed, it's the only thing that ever has.

Margaret Mead

CONCERN FOR THE ENVIRONMENT BEGAN DECADES BEFORE the start of the major environmental movement and the passage of environment-related laws. People who lived near heavy industry didn't like dirty air, but they believed industrial smoke was a necessary evil. No industry or community wanted to have to spend vast amounts of money cleaning up its waste, because doing so would be so expensive that companies thought it would put them out of business. And certainly no business wanted to be the first to do so, knowing that if its competitors didn't do the same, it would likely go bankrupt.

One of the first government environmental initiatives was the establishment in 1872 of Yellowstone National Park, the world's first national park. In 1891 John Muir founded the Sierra Club. And with other national parks being created, in 1916 the National Park Service was founded. The government started to directly address environmental pollution issues as early as 1948, when the Federal Water Pollution Control Act was passed. That law confirmed that water pollution was not good but said that controlling it was up to the states. The federal government's role was defined as support and help for technical research to devise methods of treatment for industrial waste. In spite of its name, the act achieved little.

THE 1960S: FROM CONSERVATION TO
ENVIRONMENTAL PROTECTION

The impact of private grassroots organizations on the struggle to en-act tougher environmental regulations cannot be overstated. National organizations such as the Sierra Club, the Izaak Walton League, the National Audubon Society, the National Wildlife Federation, and The Nature Conservancy, together with local groups such as the Save the Dunes Council and the Grand Cal Task Force, have been the vehicles by which the American public has become aware of environmental haz-ards and the possibility of progress through environmental activism and governmental legislation. These organizations were at the forefront on environmental issues and helped create the groundswell that finally con-vinced cities, states, Congress, and the international community that environmental progress was not just possible but absolutely necessary.

Rachel Carson's book *Silent Spring* was published in 1962. In it she took a critical look at the consequences of pollution. She also claimed that DDT (dichlorodiphenyltrichloroethane) and some other pesticides had been found to be carcinogenic. She maintained that the widespread use of these pesticides was killing birds and thus making the forests "si-lent." Her book caused quite a sensation. It received a positive review in the *New York Times* and portions were published in *Audubon* magazine. As expected, the chemical industry strongly criticized the book and initi-ated a vigorous campaign to defend and promote the use of pesticides.

Just a few months after the book's publication, the *CBS Reports* spe-cial "The Silent Spring of Rachel Carson" was broadcast and seen by an estimated ten to fifteen million people. The program included interviews with many experts, mostly critics, but the reaction from the public was quite positive and the program was one of the factors responsible for Congress's deciding to conduct a review of the dangers of pesticides. All this attention helped bring environmentalism to the forefront of the American consciousness.

It took ten years, but in 1972 DDT was banned for agricultural use in the United States. It is still allowed to be used in the fight against malaria. In 2001 the Stockholm Convention on Persistent Organic Pollutants issued a worldwide ban on DDT as an agricultural pesticide. As before,

DDT was not banned for malaria control until an affordable substitute could be found.

WATER QUALITY ACT OF 1965

Responding to citizen concerns about water pollution, in 1965 Congress passed the Water Quality Act, which strengthened the 1948 Water Pollution Control Act that had weakly begun the cleanup movement by setting principles that should be followed in the pursuit of water quality. One of the provisions of the 1965 act was that states were required to establish standards for water quality for interstate waters. It further authorized the Federal Water Pollution Control Administration to set standards whenever states did not. The act authorized funding for more research and development of systems for controlling combined sewer overflows (CSOs), which were causing much of the nonindustrial water pollution across the country and in the Calumet Area. But the absence of any mandate for CSO control meant that the CSO problem received little attention.[1]

EARTHRISE: A 1968 PHOTO THAT CHANGED THE WORLD

Astronaut and Gary native Frank Borman was the mission commander of Apollo 8, the first manned spacecraft that went to and orbited the moon. He and his crew were the first human beings ever to see the far side of the moon, and the first to see the entire planet Earth. The taking of this photograph was not on the astronauts' schedule, but on December 24, 1968, when the earth came into view as the spacecraft orbited the moon, all three astronauts marveled at the sight.

Frank Borman took the first-ever photo of the earth taken from the moon, but as his camera had only black-and-white film he ordered color photos taken as well. Astronaut Bill Anders took the photo now called *Earthrise*. Seeing Earth from space, fellow astronaut Jim Lovell noted that "the vast loneliness up here of the moon is awe-inspiring and it makes you realize just what you have back there on Earth."[2]

Thirty-five years later *Earthrise* was not only included in *Life* magazine's book *100 Photographs That Changed the World*, but it was placed in the center of the book's cover—twenty times larger than any other photograph. Galen Rowell, a well-known "adventure" photographer, is quoted in the book as saying that *Earthrise* was "the most influential environmental photograph ever taken."[3] A version of the photo was also reproduced on a first-class stamp in 1969. The photograph helped people appreciate just how self-contained our planet is, and its publication has been described, perhaps overenthusiastically, as the beginning of the environmental movement.[4]

SCOOP JACKSON, LYNTON KEITH CALDWELL, AND NEPA

Washington senator Henry M. "Scoop" Jackson was described by the *Seattle Times* as "one of the first environmentalists in a state where the vast majority of citizens now identify themselves as environmentalists."[5] In 1963 Jackson became chair of the Senate's Interior Committee, where he was largely responsible for overseeing the passage of the Wilderness Act of 1964 and the National Historic Preservation Act of 1966.

During the 1960s public support grew for Congress to do something to improve the environment. By the middle of that decade environmental concerns were becoming a major issue.[6] When it was time to write environmental protection legislation, Jackson relied heavily on his Interior Committee consultant, Indiana University professor and former Hammond resident Lynton Keith Caldwell. The National Environmental Policy Act (NEPA) was heavily based on Caldwell's 1968 report, *A National Policy for the Environment*. Jackson introduced the bill in February of 1969. It was at the Senate hearing for this bill in April that Keith Caldwell suggested requiring agencies to submit an evaluation of the environmental impact of proposed projects. This all-important "environmental impact statement," which would have outside party input, and would require comments on that input, was then made a prerequisite for approval to be given for major projects.

Eleven years later, Richard Liroff, a senior associate with the Conservation Foundation, called NEPA "the Magna Carta of the environmental movement."[7] With its requirement for public input and Caldwell's environmental impact statement, NEPA has become one of the most emulated pieces of legislation in the world, having been copied by more than ninety countries worldwide as well as by the World Bank and the European Economic Community.

Professor Keith Caldwell with Al Rosenthal, looking at Caldwell's
recently published bibliography, *Science, Technology, and Public
Policy*, March 1968. *Indiana University Archives.*

While Congress was debating the NEPA bill, in June 1969 there was
a fire on (not just next to) northeast Ohio's Cuyahoga River, one of the
most polluted rivers in the United States. Actually an earlier fire in 1952
had caused more damage than this one, but *Time* magazine's 1969 article using the 1952 photos made water pollution a national topic. (*Time*
used the fire photos from 1952 because the 1969 fire was extinguished
before reporters and photographers arrived.) The fire and the uproar
about water pollution seemed to convince undecided members of Congress that environmental protection was vitally important to the nation's future. The bipartisan support then was outstanding. The Senate
unanimously passed the National Environmental Policy Act on July
10, 1969. The House of Representatives passed it by a vote of 372–15 on
September 22.

1970: A WATERSHED YEAR

The National Environmental Policy Act was signed by President Richard Nixon on January 1.

The Clean Air Act of 1970 was passed by Congress.

The first Earth Day was celebrated nationwide on April 22.

The Lake Michigan Federation, now the Alliance for the Great Lakes, was formed on May 2.

The recycling symbol was designed by Gary Dean Anderson.

The Environmental Protection Agency was established by Congress on December 2.

A Clean Air Law was passed by the Gary City Council on December 15. It was Gary's first. Perhaps emboldened by the events earlier in the year, on that night 350 disgruntled Gary residents filled the Gary City Council chambers. On the agenda that night was an amendment to the city's air pollution ordinance that, for the first time, would require U.S. Steel to curb air emissions from its coke ovens. Apparently there was some doubt as to whether or not it would pass.

Urban historian Andrew Hurley described the meeting: The great majority of the audience, fed up with years of filthy air, favored the amendment's passage. The only person present who spoke against it was a steel company vice president who rhetorically asked the audience, "Is this action of sufficient importance to warrant the imposition of an insurmountable technological and economic burden on much needed steel production?" The crowd responded with a loud "Yes!" The council unanimously approved the amendment.[8]

Two weeks later, on December 31, President Nixon signed the Clean Air Act of 1970. In his remarks at the signing, he said, "I think that 1970 will be known as the year of the beginning, in which we really began to move on the problems of clean air and clean water and open spaces for the future generations of America."[9]

The Clean Air Acts

Although the Clean Air Act of 1970 is the best known law that Congress passed concerning air quality, it was not the first. The first national-level air pollution legislation was the Air Pollution Control Act of 1955, which funded research to determine the scope and sources of air pollution. In 1963 Congress passed the first Clean Air Act, which authorized a national program to address problems related to air pollution. The Air Quality Act of 1967 then expanded research activities and importantly set a precedent by, for the first time, authorizing enforcement procedures for air pollution problems involving the interstate transport of pollutants.[10]

However, it was the enactment of the Clean Air Act of 1970 that created a truly effective national program to clean the air, and its effects were greatly felt in heavily polluted areas such as Northwest Indiana. This legislation authorized the establishment of National Ambient Air Quality Standards (NAAQS), National Emission Standards for Hazardous Air Pollutants (NESHAPs), requirements for control of motor vehicle emissions, and requirements for State Implementation Plans to achieve its standards.[11] The Clean Air Act was revised in 1990, giving the EPA greater authority to enforce air-quality regulations.

Environmental Protection Agency, 1970

The United States Environmental Protection Agency (EPA) was established on December 2, 1970, in order to consolidate in one central agency many of the varied programs related to research, standards, monitoring, and enforcement activities and thus ensure environmental protection. Several months earlier, President Nixon had called for the creation of "a strong, independent agency" that could "make a coordinated attack on the pollutants which debase the air we breathe, the water we drink, and the land that grows our food."[12] Congress agreed that a new agency was necessary to carry out the provisions of the recently passed Clean Air Act. Its authority was strengthened with the passage of the Clean Water Act in 1972.

The EPA regulates automobile emissions and has banned the use of DDT for most operations. It promotes recycling and the cleanup of toxic waste, protects the ozone layer, and revitalizes brownfields. Its efforts over more than forty-five years have resulted in cleaner air and water, and better-protected land.[13] One of the agency's earliest accomplishments was the removal of lead from gasoline.

Many of the recommendations and requirements set by the EPA were contested by the energy and manufacturing industries as being too ex-

William Ruckelshaus, Indiana native
and the first and fifth administrator
of the EPA. Ruckelshaus set up
the agency, hired its first staff,
and decided its priorities.
US Environmental Protection Agency.

pensive. Deadlines were missed. Opposition to any requirements at all grew. But a number of environmental disasters, such as the one at Three Mile Island, gave the EPA more support, and it came to be seen by many as a health protection agency.[14] Its budget grew as Congress adopted hazardous-waste and groundwater regulations. In 1979, the "Valley of the Drums" problem near Louisville, Kentucky, where thousands of leaky, rusted barrels containing hazardous wastes were found, led to the establishment of the Superfund program.

The Region 5 office of the EPA is in Chicago. It serves Illinois, Indiana, Michigan, Minnesota, Ohio, and Wisconsin.

First Earth Day, 1970

In 1969, Gaylord Nelson, a senator from Wisconsin and the founder of Earth Day, was known as the leading environmentalist in the US Senate. In the previous ten years he had visited twenty-five states and given hundreds of speeches on the need for environmental protection and improvement. He knew that there was widespread concern about the environment, and decided that what was needed was a dramatic, nationwide demonstration.

As the Senate was considering the National Environmental Protection Act, Nelson decided to organize a nationwide "teach-in." With bipartisan support, his organization Environmental Teach-In Inc., ran ads in newspapers and coordinated many of the events. Acceptance of the idea spread quickly. Governors and mayors issued proclamations; college campuses and environmental groups scheduled activities. A full-page ad in the *New York Times* on January 1 used the name "Earth Day." And on April 22, an estimated twenty million people attended rallies, concerts, marches, work camps, and teach-ins.[15] Many wrote letters and signed petitions promoting environmental change. In Washington, DC, ten thousand people heard folk music from Pete Seeger and Phil Ochs. Afterward, Earth Day organizers reported that two thousand colleges and ten thousand schools had held activities that day. The National Education Association estimated that in all, ten million schoolchildren took part in Earth Day activities.[16]

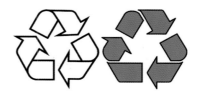

Two versions of the recycling symbol designed in 1970 by then college student Gary Dean Anderson. The symbol is used on materials that can be recycled, and not, as is believed by some, to indicate that the item was itself made from recycled material.

Events in Northwest Indiana included a weeklong teach-in at the IU Northwest campus in Gary. Individual sessions on air, water, and land pollution, and one session on ecology and noise pollution preceded a presentation by the Indiana secretary of state on Friday of that week.

Recycling and the Recycling Symbol, 1970

Recycling didn't begin in the 1960s—people have been doing it for millennia. U.S. Steel's coke plant recycled long before the term became popular. For example, the company used fans to draw fumes directly from its coke ovens so that certain gases could be sold to chemical companies for the manufacture of perfumes, aspirin and other drugs, fertilizers, and paints, as well as DDT. Then, because the remaining gas contained large amounts of carbon, it was sent back to the ovens and used as fuel.[17]

However, recycling became a new practice for many people in the latter part of the twentieth century because so many household items had by then become "disposable." Thick newspapers with dozens of inserts, catalogs, magazines, razors, batteries, bottles, cans and jars, plastic spoons and forks, plastic and paper plates and cups, paper towels and napkins, paper bags, and plastic bags were cheap and disposable. To this list could be added telephones, toasters, refrigerators, and other household appliances, and even automobiles, which, of course, aren't discarded after one use, but are still discarded after many uses.

ALLIANCE FOR THE GREAT LAKES

ENSURING A LIVING RESOURCE FOR ALL GENERATIONS

The Lake Michigan Federation / Alliance for the Great Lakes

The Alliance for the Great Lakes is the oldest grassroots environmental organization in North America dedicated to protecting all five Great Lakes. It was formed in Chicago on May 2, 1970, by a group of environmental activists from all four states that border Lake Michigan. Organized by Openlands Project staff member Lee Botts, it was originally called the Lake Michigan Federation but later expanded its focus to include all the Great Lakes, and thus adopted its current name. In 1974 the federation led the successful national effort to lobby Congress to pass the Toxic Substances Control Act, which banned DDT for most uses in the United States

One of the alliance's very successful projects is its annual beach cleanup in which more than ten thousand volunteers scour beaches for litter as well as monitor beach health. At the time of this writing, the alliance was working to build a consensus to ban microplastics in cosmetics.

The Clean Water Act

The Clean Water Act, passed by Congress in 1972, was actually an amendment to the earlier and much weaker Federal Water Pollution Control Act of 1948. It was summarily vetoed by President Nixon, not because Nixon disagreed with the intent of the law, but because he was unwilling to spend the money needed to accomplish its goals. However, Congress was able to override the veto with strong bipartisan support. The act established the structure for regulating quality standards for surface waters and for regulating pollutants entering into the nation's waters. It also made it unlawful to discharge pollutants from any point source into navigable waters (which the federal government could control), unless a permit was obtained, and to help deal with these pollutants it provided funds for the construction of local sewage treatment plants. In addition,

President Nixon and Prime Minister Trudeau signing the Great Lakes
Water Quality Agreement of 1972. *International Joint Commission.*

the act also established the Construction Grants Program to assist communities in improving their wastewater infrastructure.[18]

The Great Lakes Water Quality Agreements

On April 15, 1972, Prime Minister Pierre Trudeau and President Nixon
signed the first Great Lakes Water Quality Agreement, in which Canada
and the United States committed to improving the water quality of the
Great Lakes.[19] In essence, the two countries agreed to clean up the Great
Lakes, which contain 90 percent of US fresh water and supply drinking water to nearly twenty-five million people. Six years later, the 1978
agreement established a list of toxic chemicals for priority action and a
philosophy of "zero discharge" of harmful pollutants.[20]

Then, in a shared effort to reduce pollutants in the most polluted parts
of the Great Lakes, the agreement was amended in 1987 to ensure that
each country would commit to implementing "Remedial Action Plans"
for forty-three "Areas of Concern" (AOCs), or regions that have "experienced environmental degradation."[21] One of those areas was the
Grand Calumet River. In 2012 the United States and Canada updated
the accord again.

The effect of a north wind: smoke (particulate matter) as well as invisible matter gets blown over residential neighborhoods, June 1958. East Chicago's Jeorse Park is in the foreground. *Calumet Regional Archives.*

The Road to Cleaner Air   9

A MAJOR DIFFERENCE BETWEEN THE EFFORTS ONE HUNDRED years ago and those today is that we no longer just want parks to be established so that we can retreat to them to get fresh air—we want the cities' air to be fresh as well! Air pollution comes from both natural and human-related sources. Natural sources include volcanic eruptions and sometimes wind-blown dust. Human-related sources include moving vehicles and stationary sources of all sizes, from large factories and mills to small dry cleaners.

Fifty years ago the air pollution problem in Northwest Indiana was obvious to even the casual visitor. Particulate matter pollution in Lake County was for decades the worst in the state. In the 1970s and '80s, the county's total suspended particulates frequently exceeded health standards by significant margins. One event that significantly reduced particulate matter in the air was the closure of Inland Steel's heavily polluting coke ovens.

Even as recently as 2002, ozone was still difficult to reduce. Lake County was within the Chicago/Northwest Indiana severe nonattainment area for ozone, that is, an area in which the ozone level is significantly higher than what is considered acceptable by EPA. The highest levels were often recorded during extremely hot summers.

Several process changes in the county have led to a significant reduction of lead in the air. The federal phaseout of lead in gasoline has reduced lead concentrations. The two Hammond Lead plants both upgraded their ventilation and filtration systems, going beyond what was required by law. The East Chicago U.S. Smelter and Lead Refinery closed in 1985.

Its extremely contaminated seventy-nine-acre plant is now a Superfund site. IDEM has not found the air to exceed the lead standards since 1986.

Cars, trucks, buses, and other modes of transportation are a big part of the air pollution puzzle, especially as the number of vehicles in Northwest Indiana grows. Because vehicles emit fewer pollutants when working properly, for a number of years Lake and Porter Counties have had a vehicle inspection and maintenance program. This program is intended to ensure that cars are operating with pollution controls in place and in good condition.

The quality of the air in the United States, in Indiana, and in the Calumet Area has significantly improved over the last thirty years and indeed it is much better now than it was even ten years ago. IDEM commissioner Tom Easterly has reported that in 2009, for the first time since the federal air quality standards were developed in the 1970s, all Indiana residents were breathing air that met the standards.[1] This was a significant accomplishment, considering that four years earlier the state had twenty-four counties and townships that were not meeting the ozone standard and seventeen counties and townships that were not meeting the standard for fine particulate matter.

IDEM operates a year-round monitoring network that is more extensive than those of most other states. In 2015 all monitors in Northwest Indiana were meeting standards for carbon monoxide, nitrogen dioxide, and sulfur dioxide. That same year, measured air quality in Lake and Porter Counties was better than the federal air quality standards, though IDEM states that those counties are still designated as nonattainment for ozone as a result of Chicago's air quality.[2] Measured air quality in LaPorte County is better than the federal standards for all pollutants except ozone.

Of course, there is more work to be done—for instance, East Chicago's levels of lead and LaPorte County's ground ozone levels are higher than desired. But even they are much better than they were thirty years ago.

Although there is still a yearly increase in car and truck miles driven in Northwest Indiana, the amount of pollutants emitted by these vehicles is decreasing each year. This decrease in emissions can be attributed to a variety of national clean air programs, such as emission standards for vehicles and gasoline, the Acid Rain Program, the Heavy-Duty Highway Diesel Program, and the Clean Air Nonroad Diesel Rule. In addition, most Northwest Indiana communities have regulations concerning the open burning of trash and leaves. IDEM expects that this downward trend in harmful emissions will continue as existing clean air programs continue and new programs and recently adopted state rules are implemented.

The Clean Air Act of 1970 set the stage for national, state, and local efforts to enhance and protect air quality. Under the act, the EPA established standards (called the National Ambient Air Quality Standards, or NAAQS) for harmful pollutants, and in cooperation with state and local governments, the agency is also responsible for making sure that those standards are met. In Northwest Indiana, this monitoring is done by the EPA and by IDEM. Each state containing nonattainment areas must develop a plan for cleaning the air in those areas.

When the Clean Air Act was revised in 1990 it gave the EPA greater authority to enforce air quality regulations. Realizing that efforts to reduce air pollution were often very expensive, the act also emphasized the importance of using cost-effective approaches in reducing air pollution. In 2010, on the fortieth anniversary of the 1970 Clean Air Act, the EPA noted the following:

- For more than forty years, the Clean Air Act has reduced pollution even as the US economy has grown.
- Americans now face lower risks of premature death and other serious health effects.
- Environmental damage from air pollution is reduced.
- The value of Clean Air Act health benefits far exceeds the costs of reducing pollution.
- The act has prompted the deployment of clean technologies, and has provided an impetus for technology innovations that reduce emissions and control costs.
- New cars, trucks, and nonroad engines use state-of-the-art emission control technologies.
- New plants and factories install modern pollution-control technology.

. Actions to protect the ozone layer are saving millions of people from skin cancers and cataracts.

. The scenic vistas in our national parks are clearer due to reductions in pollution-caused haze.

. The EPA has taken initial steps to limit emissions that cause climate change and ocean acidification.[3]

In the past, the federal, state, and local governments have had to cite many industries and municipalities and force some cleanups. A big change today is that many companies and citizens are cleaning things up simply because they know it's the right thing to do.

PROGRESS IN LOWERING CRITERIA AIR POLLUTANTS

The EPA has identified six principal (or criteria) pollutants, all of which have severe consequences for both human health and the environment: carbon monoxide, lead, nitrogen dioxide, ozone (or smog), particulate matter, and sulfur dioxide.

Carbon monoxide (CO) is a colorless, odorless gas emitted during incomplete combustion. The majority of CO emissions comes from vehicles such as cars and trucks. However, in many urban areas, including the Calumet Area, industrial processes are another major source. Inhaling CO results in a reduction of oxygen in body tissues and organs (such as the heart and brain). At very high levels, CO is fatal. There was an 85 percent reduction in CO from 1980 to 2014.

The major sources of lead emissions have typically been motor vehicle fuels and industrial sources. Lead adversely affects kidneys and the nervous, immune, reproductive, developmental, and cardiovascular systems. Children are especially sensitive to lead, which may contribute to behavioral problems, learning deficits, and lowered IQ.[4] There was a 98 percent reduction in lead from 1980 to 2014.

Nitrogen dioxide (NO_2) and nitrogen oxide (NO) are two of the several oxides of nitrogen (NO_x). The EPA monitors NO_2 as an indicator for the others. NO_2 quickly forms from emissions from cars, trucks, buses, power plants, and off-road equipment. It is linked with a number of adverse effects on the respiratory system and also contributes to the formation of ground-level ozone.[5] There was a 60 percent reduction in NO_2 from 1980 to 2014.

Ozone is an essential component of the upper atmosphere but is harmful near the ground. Ground-level ozone is created in the presence of sunlight by chemical reactions between nitrogen oxides (NOx) and volatile organic compounds (VOCs). Factories and utilities, vehicle exhaust, and chemical solvents are major sources of NOx and VOCs. There was a 33 percent reduction in ground-level ozone from 1980 to 2014.

Nationally, the largest source of sulfur dioxide (SO_2) emissions is the burning of fossil fuels such as coal. Other sources include the burning of high-sulfur fuels by vehicles such as ships and locomotives. SO_2 is associated with adverse effects on the respiratory system, especially in children, the elderly, and those with asthma. There was an 80 percent reduction in SO_2 from 1980 to 2014.

Particulate matter (PM) is tiny solid or liquid particles in the air. Some particles are large enough or dark enough to be seen as smoke. Others are so small that they can be detected only with a microscope. PM affects breathing and respiratory systems and can cause cancer, lung tissue damage, and premature death. Children, the elderly, and those with chronic lung disease, influenza, or asthma are especially sensitive to particulate matter in the air.[6] $PM2.5$ and $PM10$ particles are 2.5 and 10 micrometers across, respectively. There was a 36 percent reduction in $PM10$ from 1990 to 2014, and a 35 percent reduction in $PM2.5$ from 2000 to 2014.

CORPORATE EFFORTS TO REDUCE AIR POLLUTION

Northwest Indiana's largest industrial corporations have taken significant steps in decreasing their polluting emissions. The following details several of these efforts and the positive effects they have had on the region's environment.

BP

BP, like Amoco before it, has significantly reduced its emissions. It has thirty continuous emission monitors throughout its refinery. If emissions increase, corrective actions can be taken right away. As part of a voluntary agreement with the EPA, BP has reduced hydrocarbon emissions through programs such as enhanced valve monitoring, enhanced pump monitoring and repair, and the reduction of hydrocarbon flaring. Beginning in 2002, SO2 and NOx emissions were reduced through additives used at its Fluidized Catalytic Cracking units, the use of a new Selective Catalytic Reduction system, the elimination of burning oil in all of its refinery heaters, and the installation of Ultra Low NOx burners at selected power station boilers.[7]

As part of its most recent expansion program the company invested more than $1 billion in additional environmental enhancements, including improvements in the treatment of wastewaters, reduction of air emissions, and systems that take sulfur out of gasoline and diesel fuels. In addition, equipment at the refinery has been replaced with newer technologies, and innovative emission controls have been installed on new and existing units.[8]

Inland Steel Company

Inland Steel celebrated its centennial in 1993. In its anniversary history booklet, Jack Morris noted that the company had used flue gas scrubbers on its blast furnaces for several decades, but in 1965 it still captured only 40 percent of open-hearth dust. He also pointed out that it wasn't easy to install pollution-control equipment on existing equipment. For instance, during the 1970s the company installed two experimental coke batteries with unique pipeline charging systems to greatly reduce air emissions, but that process proved to be far too expensive to maintain and was soon discontinued. Altogether, from 1950 to 1993 the company invested more than $600 million in environmental controls, two-thirds of which was spent in just the ten years following the passage of the Clean Air Act.[9]

ArcelorMittal

ArcelorMittal USA has owned the former Inland plants since 2006 and the Bethlehem plant at Burns Harbor since 2007. The company affirms on its website that it believes it has a responsibility to protect local ecosystems in the areas where its plants are located, and that it is actively pursuing reductions of air emissions through its energy-efficiency projects and by using natural gas for its blast furnace operations, as gas is better both environmentally and economically.[10]

In addition, the ArcelorMittal policy, as stated on its website, is to lessen climate change by reducing CO_2 emissions. Specifically, it states that by 2020 it will cut its CO_2 emissions by 8 percent per ton of steel produced. Another goal is to address climate change through more energy-efficient facilities. By increasing efficiency as well as the strength and durability of steel products, its studies show that there will be significant CO_2 savings from the use of steel in the automotive and other industries.[11]

Northern Indiana Public Service Company (NIPSCO)

NIPSCO has done much to improve the air quality of Northwest Indiana and is still investing in additional pollution-control devices. Since 1990, the company has invested $350 million in new pollution controls and has reduced SO_2 and NOx emissions by 70 percent. Since just 2005, NIPSCO has seen a reduction in greenhouse gas emissions of 18 percent, which the company points out is well above the national average of 10 percent for all industries.[12] NIPSCO was pleased to report that in part due to its investments in environmental controls, Northwest Indiana in 2010 was finally designated an attainment area, for the first time since the passage of the 1990 Clean Air Act.[13]

By 2018, NIPSCO will have completed another $800 million investment in additional environmental controls. For example, NIPSCO's NOx emissions will be 35 percent below current rates, SO_2 emissions will be 80 percent below current rates, and other benefits, such as reduced fleet vehicle emissions and improved air quality monitoring, will have been achieved.[14] After these additional investments, NIPSCO expects that its coal fleets will be among the cleanest in Indiana, and that the company will be fully scrubbed by 2016.[15]

NIPSCO notes that two of the factors that have allowed the company to reduce its carbon footprint are the upgrading of its existing coal fleets to make them more efficient, resulting in less fuel being required to generate the same amount of electricity, and the addition of NIPSCO's customer programs, such as its energy-efficiency programs.[16]

United States Steel

The U.S. Steel website states that one of the core values of the company is environmental stewardship, and that it articulates that value through four basic principles:

· Compliance with environmental laws and regulations
· Continuous improvement in environmental and resource management
· Continued reduction of emissions
· Community partnerships to protect and preserve natural resources[17]

In 2011 U.S. Steel started using compressed natural gas–powered vehicles. Up north, the company partnered with Minnesota Power to create northeastern Minnesota's first commercial wind center, which uses ten wind turbines to produce twenty-five megawatts of power.

U.S. Steel notes on its website that the making of steel uses less energy and emits less CO_2 per ton than either aluminum or magnesium. In addition, steel, with more than eighty million tons of it recycled each year, is now the world's most recycled metal. Nevertheless, in the last twenty-five years, through the adoption of thin-slab, flat-rolled production using electric arc furnaces, along with other new technologies in steelmaking, the steel industry has reduced the amount of energy required to produce its product by 31 percent and has reduced CO_2 emissions by 36 percent.

U.S. Steel has invested more than $400 million in air and water quality-control systems. Air-quality systems include venturi scrubbers and electrostatic precipitators. The company regularly cleans its roads within the plant in order to reduce fugitive dust. Its efforts have reduced air emissions by more than 95 percent. One of its new programs, Continuous Improvement to the Environment (CITE), is a proactive effort to teach employees how to make environmental activity part of their "continuous job responsibility." The company is proud that Gary Works' environment is such that it has attracted a pair of peregrine falcons that have nested on a bridge support within the plant facility.[18]

LOOKING TO THE FUTURE

Indiana is still making progress in meeting and exceeding clean air standards. In 2005, thirteen Indiana counties had air quality that did not measure up to the standards. There is much to celebrate, but vigilance is still necessary. There will always be work to be done.

There are some educated well-meaning people who believe that today's standards are too lax, and others who believe they are too stringent. Science will surely discover more about pollutants and their effects on humans and the environment. Technologies for monitoring and cleaning the air will advance. On occasion, accidents may occur or there may be deliberate actions by some that foul the air. Consequently, monitoring must continue so that when breaches occur they may be found quickly and remediated as soon as possible. But we can all still rejoice over the progress made since the National Environmental Policy Act became the law of the land back in 1970.

The Grand Calumet River at Roxana Marsh, October 2011. One can see in the river two hydraulic dredges in the process of removing contaminated sediments. Note how the residential area to the left was screened by trees so that residents didn't have to see the then-polluted river from their homes. The greenish-gray material on both sides of the river is phragmites, an invasive, nonnative plant. See chapter 16 for a description of Roxana Marsh's environmental transformation. *Tetra Tech for the US Environmental Protection Agency.*

The Road to Cleaner Water 10

Just as it is particularly blessed, Northwest Indiana is particularly challenged to achieve a cleaner, safer, richer environment and a sustainable balance between nature and the built environment.

One Region, Northwest Indiana Profile

ACCORDING TO THE INDIANA DEPARTMENT OF ENVIRONMEN-tal Management (IDEM), the state has 63,130 miles of rivers, streams, ditches, and drainage ways within its borders.[1] Fifty years ago many of them were in bad shape. The worst was the Grand Calumet River, a waterway that was in effect a free disposal system for municipal and industrial waste.[2]

Today, it's a different story. Hoosier waters are cleaner and safer. The cleanup has been slow, but steady. Attitudes are different as well. Rivers are no longer thought of as natural sewers. Citizens now become upset when municipalities or industries even accidentally pour pollutants into rivers and streams.

Although early laws made it illegal to dump pollutants into lakes and streams, federal legislation in the 1970s strengthened these rules and finally made enforcement a reality. But at that time few municipalities, businesses, or industries felt that they could afford to clean up their effluent, especially if no one else did. Cleanup would be and has been expensive. Removing pollutants from effluent has made a big difference in water quality, but it didn't remove the contaminated sediments deposited over a century of unregulated disposal. That took more effort.

DRAINAGE SWALES AND SEWERS

In many communities, aboveground drainage ditches or swales were often the first public improvement designed to quickly move waters from

U.S. Steel workers digging a trench for the company's private
sewer system, July 1906. *Calumet Regional Archives.*

either the city or potential farmland into a nearby stream or river. Scores
of these ditches still do their work in the Calumet Area; Burns Ditch
(now called Burns Waterway) is certainly the largest. Hart Ditch drains
parts of Schererville, Dyer, Munster, and Highland. Cady Marsh Ditch
drains parts of southern Gary as well as Griffith and Highland. Schoon
Ditch drains Munster south of Ridge Road. Kintzele Ditch drains parts
of Beverly Shores and Pines.

As cities grew, many of these ditches were replaced by underground
sewers. Though the sewers were originally intended to drain just storm
water, folks soon realized that they could easily direct their own sewage into these drains and thus gradually they became both sanitary
and storm sewers. Soon all, or nearly all, sewers carried both waste
and storm water. At the lower end of these drains, the contents were
dumped into a stream or river. There the wastewater was diluted by the
river water.

Gary town board members inspecting the city's new sewers,
August 1908. *Calumet Regional Archives.*

By the beginning of the twentieth century, it was expected that cities
would have sewers. Gary, the youngest of the four Lake County lakeside
cities, installed sewers as the city was first developed in 1906. At the same
time, U.S. Steel installed its own private sewer system as it built its lakeside plant in Gary. Most of its waste was dumped into the Grand Calumet
River, while some was directed north into Lake Michigan.

In these early days, of course, sewage wasn't treated, but that wasn't
considered a problem. "Dilution as the solution to pollution" was widely
accepted. The disposing of sanitary sewage into rivers and streams diluted the sewage, and that method continued to be the manner in which
American cities handled their wastewater until criticism of this method
became common around the 1910s–1920s.[3]

When Chicago in 1900 reversed the flow of the Chicago River, so that
the sewage dumped into that river would then flow westward, away from
Lake Michigan, the city did not install any sewage treatments. It simply
diluted the sewage with waters coming from the lake and sent them

downstream toward the Illinois and Mississippi Rivers. The communities downstream, such as Lockport and St. Louis, weren't pleased, and a court-ordered injunction was threatened but didn't occur. According to the *Chicago Tribune*, downstream communities didn't suffer.[4]

Until 1917 most (if not all) Indiana communities that had sewers dumped all their wastewater directly and untreated into nearby streams (downstream of the town itself, of course). That year, the Indiana General Assembly passed the Sanitary District Act, which authorized local communities to create sanitary districts for the purpose of providing sanitary services. Even then, Indiana communities delayed establishing this additional taxing authority. Gary's and Hammond's districts, for instance, weren't established until 1938.

By this time, Chicago and Milwaukee were both removing the solid material from their sanitary sewers, drying it, and converting it into fertilizer that was sold to midwestern farmers.[5] Many Calumet Area plants do the same today. But now they do more than that.

Gradually all the Calumet Area cities established their own districts, or made arrangements with nearby cities to process their wastewater. (Sewage from Merrillville, Lake Station, New Chicago, and Hobart all goes to Gary's plant. Griffith's, Highland's, and Whiting's goes to the Hammond/Munster plant. Porter's waste is handled by Chesterton.) Over the years, the various plants have expanded and modernized their operations from the simple primary treatment, which removed solids and oils, to secondary and tertiary treatments. Those operations that now have tertiary treatment remove 99 percent of harmful pollutants from the water that they treat.

A significant amount of pollution in rivers and lakes today, however, is caused by combined sewer overflows (CSOs). During heavy rainfalls or fast snowmelts, much more water heads into city sewers than during other times. For instance, the Hammond plant normally handles thirty million gallons of wastewater a day, but during and just after heavy rains the amount of water that enters the sewers can quickly jump to eighty-five million gallons. Because sewers and treatment plants can handle only a certain amount of sewage per hour, they cannot handle all the waste and storm water during heavy downpours.[6] At such times, wastewater is dumped into local rivers untreated.

CSO discharges include residential, commercial, and industrial effluents mixed with storm water runoff.[7] This toxic mixture poisons rivers and adversely affects the health of both people and wildlife. In the Calumet Area, these polluted discharges contribute to the contamination of drinking water as well as beach closures. All, or at least nearly all, of the communities in the Calumet Area have at some time contributed to this problem.

These situations have occurred because cities and towns originally had just one sewer system that carried both storm water and wastewater. To alleviate this situation, beginning in the 1960s many communities began requiring new subdivisions to install separate storm and wastewater sewers. Then those communities that could afford to do so began separating their older sewers by installing a separate sewer for sanitation purposes.

As a result of these efforts, the number of CSOs has decreased dramatically, but not so for many communities that have been suffering from declining populations. With fewer taxpayers, these typically older cities have had their overall budgets reduced but still have the same number of miles of sewers to maintain, and many of those sewers are more than one hundred years old. For those communities, sewer separation is nearly financially impossible.

As late as 2005, thirty-three years after the passage of the Clean Water Act, Chesterton, Crown Point, East Chicago, Gary, Hammond, Michigan City, and Valparaiso sewage treatment plants were still experiencing some combined sewer overflows, most of them in the Lake County lakefront communities. That year, those four cities experienced forty-five CSOs, with Gary and Hammond (the two largest) having the most, although their numbers paled in comparison to Elkhart's and South Bend's, whose effluent enters Lake Michigan via the Saint Joseph River.

Each of those cities has been directed by the EPA and IDEM to find ways to eliminate CSOs. Hammond, for instance, has built an enormous thirty-three-million-gallon dual cell storage basin where combined sewer waters can be stored until a storm passes. CSOs now discharge into the first cell, which may overflow to the second cell if and when the first cell fills up. On very rare occasions when rainfall continues and the second basin fills, only then do waters overflow into the river, but what gets into the river at that point is remarkably cleaner than a direct CSO discharge would have been.[8] The basins were put in service on September 26, 2014, and have reduced CSO discharges from three of Hammond's largest pumping stations by more than 90 percent.[9] The district also installed larger dewatering pumps at the two off-site pumping stations (two pumps at three million gallons a day each per station), which dewater the effluent during dry weather. This then flows directly to the treatment plant.

INDUSTRIAL POLLUTANTS

The other major source of pollutants in Calumet Area waters was industry. When the steel mills, refineries, and other heavy industries located in Northwest Indiana, there were no regulations about dumping pollutants into rivers. There was also no knowledge at that time about the ill effects of these industrial pollutants. Some "experts" even claimed that all this industrial effluent helped dilute the sanitary wastes in these streams and rivers and thus contributed to the health of the region.

Many of the pollutants dumped into the Grand Calumet River were carried into Lake Michigan. Others settled to the bottom of the river or the Indiana Harbor Ship Canal and just remained there. In 1996 the EPA estimated that the river and the canal contained five to ten million cubic yards of contaminated sediments. This huge amount was due in part to the fact that the Ship Canal had not been dredged since 1972 when it became difficult to find a location to place these hazardous sediments. It was estimated at that time that the waters in the canal carried 150,000 cubic yards of these sediments annually into Lake Michigan.

In addition to all this, millions of gallons of petroleum were floating atop the groundwater of the industrial portion of Northwest Indiana. There were seven hazardous Superfund sites, hundreds of other sites that required cleanup, and numerous leaking underground storage tanks, all of which allowed contaminated water to seep into area rivers.[10]

EFFECTS OF THE CLEAN WATER ACTS

The first major effort to clean area waters was made in the 1960s by the Calumet Enforcement Conference created by requirements in the earlier 1948 Federal Water Pollution Control Act.[11] It and the four-state Lake

Michigan Enforcement Conference studied the waters of Lake Michigan and the Calumet River and then set up timetables to clean up their pollution. The conference estimated that the cost of doing so would run anywhere from $2 to $10 billion. This was money that it did not have.

The Clean Air Act in 1970 and the Clean Water Act in 1972 limited the chemical pollutants that industries and municipalities could release into the air and water, but they didn't set into action any policy for cleaning up already polluted rivers and lakes. At the time the Clean Water Act passed, two-thirds of the nation's waters were unfit for fishing or swimming, raw sewage nationwide was being dumped directly into American rivers, and most dramatically, the thick oil slicks on the Cuyahoga River in Ohio had several times caught fire.

After sixty to seventy years of operating with few governmental regulations regarding the emission of pollutants, it was not easy for businesses and industries to quickly clean up their acts. Change didn't come easily at first because there were few experts in the industrial world who fully understood the seriousness of problems associated with air and water pollution. Folks used to think it was OK to throw their waste into the water. It was a common, if not universal, practice. Industries didn't have environmental specialists on board as they do now. So today, there's a better understanding within industries of why certain standards are in place. As a result, it's now common for industries to go beyond minimum standards. Management recognizes the need for sustainability for the long-term health of the company.

The effects of the Clean Water Act were slow but steady. New industries were formed and new jobs created, all related to the effort to remove pollution from American waterways and to keep them clean. Many industries and municipalities were cited for not meeting the new clean water requirements. New wastewater treatment plants were built, old treatment plants modernized, and old sewers replaced, and many communities split their combined sewers into separate storm and sanitary systems.

All integrated steel mills use a large amount of water on a daily basis. According to Jack Morris, Inland Steel's director of corporate communications in 1993, the Indiana Harbor Works used as much water each day as the entire city of Chicago. He called steel making a "very dirty industry,"[12] and noted that the 1970s brought increasing public pressure for cleaning up its operation and eliminating harmful pollutants from the air and water. As mentioned earlier, the company installed $400 million in environmental-control equipment in the decade after the Clean Air Act and Clean Water Act were enacted by Congress.

The other heavy industries, as well as smaller businesses in the area, all had to get permits to dump effluent into area streams. With the permits, companies were given deadlines by which cleanup must occur or fines would be assessed. Of course, some businesses complied with the regulations more quickly than others.

THE GRAND CAL TASK FORCE

The Grand Calumet River is not a tourist destination. Visitors don't (yet) plan picnics on its banks. Residents who live nearby don't fish or swim in it. Many citizens are not even aware of the river. Nevertheless, a group of local residents steadily advocated for the river to be cleaned and its natural habitats restored. For decades, they operated as the Grand Calumet Task Force, a nonprofit grassroots environmental organization dedicated to the cleaning up, restoration, and protection of the Grand Calumet River, the Indiana Harbor Ship Canal, and the surrounding urban ecosystem.

The task force was organized in 1981 with help from the Chicago-based Lake Michigan Federation. Its membership included some representatives from local environmental groups, but the main corps was a racially and economically diverse group of community-minded Northwest Indiana residents who lived near the river, were disgusted with its pollution, and were willing to do something about it. The organization researched and publicized information about the river, pressured governmental agencies and elected officials responsible for enforcing environmental standards, and promoted sustainable economic development that would provide jobs without fouling the land, air, or water.

The Grand Calumet Task Force made the citizenry aware of the impacts of air and water pollution and acted as a watchdog, receiving complaints from the public, investigating problems, and urging action by governmental officials when deemed necessary. It brought blue-collar workers into decision-making operations.

A warning sign next to the Grand Calumet River. "UNSAFE WATERS: You should not swim in these waters. You should not eat fish from these waters." *Dorreen Carey.*

Passing an outfall on a Grand Cal Task Force canoe trip. *Dorreen Carey.*

A Grand Cal Task Force Earth Day rally. *Dorreen Carey.*

The task force led the successful opposition to a plan to build an island in Lake Michigan and have it contain the dredgings from the terribly polluted Indiana Harbor Ship Canal. To help make the public aware of environmental problems, the task force organized canoe trips down the Grand Calumet River, as well as "Toxic Tours" that visited places where pollutants were entering the river and areas that would become Superfund sites. Its staff wrote informational brochures such as *Where Does the Rain Go* and *Pollution Prevention* and published a quarterly newsletter. Recognizing the problems associated with abandoned former industrial properties, it applied for and received a grant to study brownfields and then organized the Northwest Indiana Brownfield Consortium. It supported an environmental justice advocacy committee.

It didn't take long for the task force to be noticed outside the immediate area. In 1992 it was named Environmental Group of the Year by the Hoosier Environmental Council, and the Chicago Audubon

Society named executive director Dorreen Carey a "Defender of the Environment."

The task force organized Earth Day activities, wrote a detailed plan for how to clean up the river, and prepared a "Gary Riverfront Revival Plan" that included a boat launch at Bridge Street, enhancements to Ambridge Park, and a multiuse trail.[13] Later the organization played a part in finding a suitable brownfield location in East Chicago for permanent storage of the heavily contaminated Indiana Harbor Ship Canal sediments. It advocated "prosperity without poison."[14]

The task force also organized Lake Michigan beach cleanups. Carrie Sanidas, a Willowcreek Middle School teacher whose students participated in this program, noted that the children "are amazed and appalled at the stuff they find" and as a result "are far more conscientious about the time they spend on the beach."[15]

In 1986 the Grand Cal Task Force became an independent 501(c)(3) nonprofit organization. Its members worked for more than twenty years bringing attention to the river, locating sources of pollutants, and encouraging legislation to prevent pollutants from entering the river. When cleanup of the river actually began, many members felt that the goals of the group had finally been realized. Task force membership declined and the organization disbanded.

In 1987, when the Great Lakes Water Quality Agreement with Canada was amended in order to reduce pollutants in the most polluted parts of the Great Lakes, the amendment identified forty-three severely degraded areas within the Great Lakes Basin as "Areas of Concern" (AoCs). Each was selected because it had at least one impairment of fourteen designated beneficial uses.[16] One of these Areas of Concern was the Grand Calumet River and Indiana Harbor Ship Canal, which was the only AoC in North America to be impaired on all fourteen uses.

Members of the International Joint Commission (IJC) visited this area after their 1983 meeting in Indianapolis. Appalled by the condition of the Grand Calumet River and apparently duly impressed with the quality and scope of the earlier cleanup plan created by the Grand Cal Task Force, the commission in its 1987 agreement required each AoC to have a similar "Remedial Action Plan" (RAP) that would outline how the area would be brought up to standard. By signing this new agreement, each country was committed to implementing these RAPs.

GREAT LAKES LEGACY ACTS OF 2002 AND 2008

Dealing with "legacy" contamination, that is, material dumped into rivers decades or perhaps a century before the Clean Water Act was passed, was one of the biggest challenges facing river restoration. Seeing a great need to accelerate the pace of sediment remediation in the Great Lakes Areas of Concern, including the Grand Calumet River AoC, Congress enacted and President George W. Bush signed the Great Lakes Legacy Act (GLLA) in 2002. Funds were then made available for projects to remediate contaminated sediment or prevent further sediment contamination in the designated AoCs. The act required that nonfederal cost share must be at least 35 percent of the total project costs and 100 percent of operations and maintenance costs. When the program was renewed in 2008, Congress added funding for habitat restoration projects in conjunction with sediment remediation.[17]

Restoration in itself is an investment. According to the Illinois-Indiana Sea Grant Program, for every dollar spent on restoration there is a benefit of two to three dollars in increased property values, a reduction in municipality costs, and economic benefits related to tourism, recreation, and sport fishing.[18] As of 2016, the GLLA has, as part of nineteen different remediation projects, cleaned up four million cubic yards of polluted sediment. In addition, restoration of upland, shoreline, and underwater habitats has been accomplished.[19]

THE GREAT LAKES COMPACT

The success of these programs led to the ratification of the Great Lakes Compact,[20] which is designed to safeguard the health of the lakes, ban the diversion of Great Lakes water (with a few limited exceptions, including the Chicago River and the Little Calumet River from Hammond and Munster westward) to areas outside of its basin, and set standards for conservation within the basin itself.[21] The compact was approved by all eight states that border the lakes and by both houses of the US Congress, and was signed by President George W. Bush in October 2008.

REMOVING LEGACY POLLUTANTS FROM
THE RIVER AND SHIP CANAL

The Indiana Department of Environmental Management was given the responsibility for drafting the Remedial Action Plan for the river/canal system. And because the cleanup of the river would involve so many communities and stakeholders, IDEM then created the Citizens Advisory for the Remediation of the Environment (CARE) Committee, which included, and still includes, representatives from the Northwest Indiana Forum; several governmental bodies such as IDEM, the Indiana Department of Natural Resources, and the EPA; and local industries and environmental organizations.[22] A Remedial Action Plan was then created. IDEM and the US Army Corps of Engineers, partnered under the Water Resources Development Act, established the Sediment Cleanup and Restoration Alternatives Project (SCRAP), which developed a feasibility study to dredge the east and west branches of the Grand Calumet River. This study became the foundation for numerous Great Lakes Legacy Act project proposals. The CARE Committee has been involved in developing, modifying, and implementing the Remedial Action Plan since its inception,[23] and still meets to discuss progress and issues related to the plan. As more information about contaminants and remediation has become available, the plan has been fine-tuned. The goal of creating a healthy waterway has not changed.

The task of cleaning up the river and canal was simply too big to be handled as one project. Thus the waterways were divided into several sections, with each segment being addressed separately.

The upper east branch of the Grand Calumet River is in Gary and much of it borders the U.S. Steel plants. At a cost of $60 million, dredging by U.S. Steel, which began in December of 2002, has removed many hazardous pollutants such as PCBs, oil and grease, heavy metals, benzene, and cyanide.[24] More than eight hundred thousand cubic yards of sediment were removed from this five-mile section and were replaced with plastic liners covered with carbon filters and clean fill.[25] U.S. Steel completed additional dredging in 2007. The corporation worked with Natural Resource Trustees in developing an in-stream restoration plan that included riffles,[26] boulder piles, and the planting of native species along the riverbank. The project was completed in July 2010. The EPA noted in 2013 that the highest-scoring samples of macroinvertebrates in the river were found, and the health of fish had improved, in the section of the river cleaned up by U.S. Steel.[27]

The plan divided the rest of the river and the two extensions of the canal into five zones:

Zone A—The west branch: Hohman Avenue to Indianapolis
        Boulevard, Roxana Marsh
Zone B—The east branch: Kennedy Avenue to Cline Avenue
Zone C—Far west: Hohman Avenue to the state line
Zone D—Far east: Cline Avenue to U.S. Steel
Zone E—The south and west forks of the Indiana Harbor Ship
        Canal, and the river from Indianapolis Boulevard to
        Kennedy Avenue[28]

As U.S. Steel had already dredged its section of the river, that stretch was not included in the plan. Neither was the main branch of the Indiana Harbor Ship Canal, which was to be dredged by the US Army Corps of Engineers.

After many years of planning the cleanup finally got underway in 2009. Zone A, the west branch and Roxana Marsh portion, was started in June of 2011. The sediments were pumped from the river into "geotubes," where they were dehydrated. The dried sediment then was disposed of at a permitted solid waste landfill. Approximately 730,000 cubic yards of sediment, contaminated with large quantities of PCBs (polychlorinated biphenyls), PAHs (polycyclic aromatic hydrocarbons), heavy metals, and pesticides, were dredged from the river. A cap was then placed over the remaining sediment. The dredging project was completed in 2012, having restored more than eighteen acres of marsh habitat and over twenty acres of wetland and riverine habitat.

Zone B, Kennedy Avenue to Cline Avenue, comprises a 1.8-mile stretch of the river and a marsh. Dredging of this zone was completed in the spring of 2015. The project team placed a cap and a clean sand layer over the base sediment. Approximately seventy-seven acres of marsh, wetlands, and nearshore riverine habitats were restored, and alien plants (including a large mass of phragmites) were replaced by native plants.

Zone C, Hohman Avenue to the state line, includes the first portions of the river polluted back in the 1870s. Its sediments contain heavy metals, PAHs, and PCBs. Dredging of this 0.7-mile stretch of the river was initiated in late summer and early fall of 2015. Early excavations in this area had unearthed some fur that might have been from carcasses from the old 1869 slaughterhouse.

As of this writing, work had not yet started in Zones D and E. Zone D is in the feasibility study stage; the EPA partner will be the Gary Sanitary District. The East Chicago Waterway Management District is the local partner for Zone E. The plan is to split this project into six areas. Proposed treatment includes both removal and capping of contaminated sediment. Work will be done as funds become available.

INDIANA HARBOR SHIP CANAL (MAIN BRANCH)

The US Army Corps of Engineers is in charge of the navigable section of the canal, which consists of the main channel and a short section of the two forks. Because the corps had not been able to find a suitable area to place dredged sediment, until the current project began the harbor had not been dredged since 1972. Therefore there was an estimated dredging backlog of 1.8 million cubic yards of sediment. This caused two problems: the pollution of Lake Michigan and impediments to deep-draft commercial navigation. A disposal area was found on former refinery property west of the canal and dredging began in the fall of 2012.

Even as late as 2002, the results of a study by Ingersoll et al. showed that sediments from the river and canal were "among the most contaminated and toxic that have ever been reported."[29] But although it has taken several decades, things have changed for the better in the Grand Calumet River and Indiana Harbor Ship Canal. Through remediation processes, more than two million cubic yards of contaminated sediment have been removed from these connected waterways. If funding from the Great Lakes Restoration Initiative is maintained and cost-sharing partners are secured, the work on the river and canal could be finished as early as 2019.[30]

Dredging on the Indiana Harbor Ship Canal on the first day of operation, October 23, 2012. *Photo by Jessica Majchrowski, US Army Corps of Engineers.*

This formerly highly polluted area is on its way to becoming a healthy riverine habitat, thanks to the partnerships formed and the efforts of the individuals working on the Remedial Action Plan. Areas that were once contaminated and plagued by invasive species are being transformed, and the plan is helping the Grand Calumet River make progress toward being truly "grand" again.[31]

LIFE IN THE GRAND CALUMET RIVER TODAY

The amount of pollutants entering the river began to decrease after the passage of the Clean Water Act of 1972, and the health of the river did begin to improve—but very slowly. In 1985, state and local officials were pleased when a carp was found swimming in the river. The poor fish was barely alive, but it was the first fish found in the Grand Calumet River in

Hammond Sanitary District superintendent Don Woodard watches
the salmon that have come to spawn in the waters coming out of the
district's discharge pipe. *Photo by John Watkins, the* Times.

decades. Later, in the 1990s, more fish were living in the river, but it was
found that more than half of them had deformities, eroded fins, lesions,
and tumors. Statewide in Indiana, only one in ten thousand fish have
these symptoms.

By 2010, many more fish were swimming in the river, and although
they tended to look healthy, the Indiana Department of Natural Resources still advised against eating them as they often still had toxins
inside them left over from those legacy pollutants dumped into the lake
decades earlier. That year, Cameron Davis, then the president and CEO
of the Alliance for the Great Lakes, noted that the river still dumped two
hundred thousand cubic yards of sediment each year into Lake Michigan—"sediment full of PCBs, heavy metals and 'some of the nastiest,
most toxic contaminants ever.'"[32]

One of the continuing problems is that "five Superfund sites border the Grand Calumet, including almost 500 underground chemical or oil storage tanks, many of them leaking. A toxic brew of chemicals and metals makes the Grand Calumet unsafe even for human contact."[33] However, since the restoration process began, more than eight species of fish have been seen in this once-dead river. In fact, each year salmon now lay their eggs alongside the Hammond Sanitary District plant on Columbia Avenue.[34]

In 2013, fish, macroinvertebrates, and sediment at the bottom of the river were sampled at various sites across the length of the Indiana portion of the river. Preliminary results concerning fish from the upper east branch of the river showed fewer toxins when compared to testing done in 1994 and 2005, and five of the six sampled sites met IDEM benchmarks for unimpaired biotic communities.[35] Preliminary results of the macroinvertebrates study indicated that none of the sites yet met those benchmarks. In what is likely a good sign of things to come, the two highest-scoring samples were found in the portion of the river that had been dredged.

Not surprisingly, sediment samples showed that concentrations of heavy metals were highest in the undredged portions of the river. Similarly, the lowest concentrations of the thirteen most prevalent PAHs and PCBs were in the dredged eastern portions of the river that had new sediment caps installed, and in the Marquette Park lagoons.

Fish tissue sampling did indicate that PCB concentrations were still among the highest in the state of Indiana. Eight of the thirteen fish tissue samples collected from the river and the Ship Canal were above the "No Consumption" benchmark. All the sites were above the "Limited Consumption for the General Population" benchmark, but mercury concentrations were below that benchmark at all locations.

After forty years US waterways are markedly cleaner. Half of the nation's waterways now meet the clean water standards, despite the fact that the population of the country has doubled, putting increased pressure on all wastewater systems.[36]

Indiana is still making progress in meeting and exceeding clean water standards. In 2005 there were still one hundred Hoosier communities discharging raw sewage into Indiana's streams. By 2015, seventy of them had appropriate sewage treatment plants or were in the process of correcting their problems. Progress continues on the remaining thirty.

Much progress has been made in the cleanup of the Calumet Area rivers and streams. It has been a slow but thoughtful process, with input from its various stakeholders every step of the way. The most dramatic change has been in the Grand Calumet River and Indiana Harbor Ship Canal, where the water quality is much better than it was before the Clean Water Act was passed. Fish swim there now, but legacy pollutants still lie on much of the river bottom and the fish that live there are not yet safe to eat.

In 2012, James Salzman wrote in *Slate* that "the Clean Water Act stands as one of the great success stories of environmental law. Supported by Republicans and Democrats alike, the act took a completely new approach to environmental protection." The law declared that "there would be no discharge of pollutants from a point source (a pipe or ditch) into navigable waters without a permit," and "provided for billions of dollars in grants to construct and upgrade" sewage treatment plants.[37]

Pollution from nonpoint sources—runoff from city streets, residential lawns, contaminated groundwaters, and farm runoff—has been harder to control. And large cities with very limited budgets, including several in Northwest Indiana, have had difficulty making the necessary investments to manage sewage overflows during major storms.[38]

Central Beach of the Indiana Dunes National Lakeshore.
Photo by Jeff Manuszak, National Park Service.

Lake Michigan Health, Beach Closures, and Fishing

11

THE FIVE GREAT LAKES CONTAIN ABOUT 18 PERCENT OF THE world's supply of fresh water and 90 percent of the fresh water in the United States.[1] Lake Michigan's deepest section is 925 feet below the surface; it is the second largest of the Great Lakes in terms of volume and is the Calumet Area's greatest resource. But Lake Michigan has been degraded over the years by pollutants entering the lake via its tributaries as well as from the air, ships and boats plying its waters, and occasional dumping. The lake has been invaded by exotic animal species that have either swum upstream into the lake or been brought into it unintentionally by anglers dumping unused bait or water or by oceangoing ships dumping their ballast water.

Chicago and all the Northwest Indiana lakefront cities get their drinking water from the lake. Although over the years most of these cities' wastewater has been dumped, not into the lake, but into the Calumet Rivers, most of these rivers eventually flowed into the lake and thus so did the sewage. Chicago was the first city to react to this unwise action, but rather than treat its sewage (as Worcester, Massachusetts, was already doing) it dredged and excavated the upper Chicago River and made the river reverse directions, flowing out of Lake Michigan and down toward Saint Louis via the Illinois River.

Up until the early part of the twentieth century, all the lakefront cities piped raw water from the lake and distributed it to customers completely untreated. Although most bacteria are harmless and some are useful, the presence of pathogenic bacteria in water causes diseases such as diarrhea, cholera, and typhoid. It was already known by 1920 that bacteria were a cause of contagious diseases, yet for decades none of the area's water-

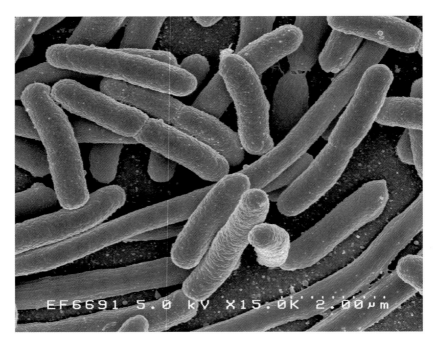

Escherichia coli bacteria as seen through an electron microscope.
National Institute of Allergy and Infectious Diseases.

works did anything to kill bacteria in the drinking water. The intake crib for Indiana cities was far enough out into the lake that people assumed that sewage from the Calumet River wouldn't affect it.

It was in 1936 that the Hammond Water Department superintendent complained about the pollution that his own city was allowing to enter Lake Michigan. Those pollutants forced his department to build a modern filtration plant during the Great Depression when money was tight. After a 1920s typhoid epidemic caused panic in Gary, many of the city's residents pressed for chlorinated water and a "modern" sewage treatment plant.[2] But it wasn't until the 1940s that the Gary Health Department began testing Lake Michigan water for harmful bacteria. When the Gary-Hobart Water Company offered to assume the role of running the waterworks, the Gary City Council agreed on the condition that it install a water filtration system at its pumping facility in Jefferson Park, which it did.

Beachgoers enjoying the Indiana Dunes State Park beach.

Escherichia coli has been a frequent pollutant in Lake Michigan waters. This bacterium can come from urban sewage dumped into rivers that flow into the lake, or it can come from rural rainwater runoff from pasture or forested lands. Even bird feces can cause the levels of *E. coli* to rise.

BEACHES

In order to improve the health of recreational waters along US coasts, including the Great Lakes, Congress in 2000 passed an amendment to the Clean Water Act called the Beaches Environmental Assessment and Coastal Health Act (BEACH Act, of course). The act was signed into law on October 10, 2000, by President Bill Clinton. This act requires each state to adopt minimum standards for recreational use of the water and to notify the public whenever beaches are not meeting those criteria. At

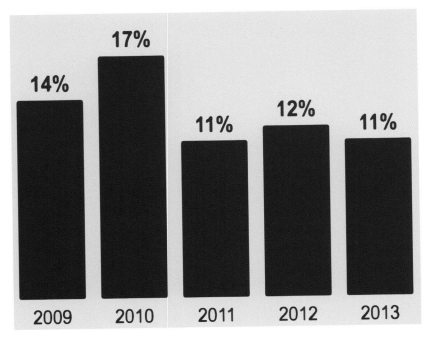

Percentage of samples exceeding the bacterial maximum for twenty-seven reporting Indiana beaches, 2009–2013. *Natural Resources Defense Council.*

the same time Congress authorized an EPA grant program to help states pay for required monitoring and public notification.[3]

Indiana's beach monitoring and notification program along the state's forty-five miles of Lake Michigan waterfront is coordinated and administered by the Indiana Department of Environmental Management (IDEM). (Monitoring is voluntary for beaches that are not on federal land.) Advisories and closures for the beaches in IDEM's program are posted on IDEM's BeachGuard website.[4] Lake Michigan beaches are safe for swimming 89 percent of the time,[5] and the worst beaches now will be much better when the combined sewer overflow problems are solved.

In 2013, thirty-two Indiana beaches were monitored for *E. coli*. The five beaches with the highest percentage of beach advisories or closures in 2013 were Jeorse Park Beach I (52 percent), Jeorse Park Beach II (40 percent), Hammond Marina East Beach (30 percent), Buffington Harbor

Beach (29 percent), and the Portage Lakefront and Riverwalk (22 percent).[6] The first four are all near the still-polluted Indiana Harbor Ship Canal; the Portage beach is adjacent to Burns Ditch. Historically, and not surprisingly, East Chicago's Jeorse Park Beach, which is the closest beach to the Ship Canal, has been Indiana's poorest-performing beach with regard to *E. coli* concentrations. This situation will vastly improve when the dredging of the Ship Canal is completed.

Although Washington Park Beach at Michigan City is adjacent to Trail Creek, the water quality there has been better than at many of the beaches to the west. Nevertheless, in the hope of reducing beach closures even further, Michigan City has recently made efforts to reduce the number of seagulls and other birds at Washington Park Beach by prohibiting the feeding of birds there and promoting the covering of trash receptacles.

One of the major pollution problems today is the occasional combined sewer overflow in the western sections of the Calumet Area. And that is being addressed by the work now being done on the Grand Calumet River and the Indiana Harbor Ship Canal.

In 2009, Congress approved the Great Lakes Restoration Initiative (GLRI), which included funds to restore the Great Lakes by cleaning up highly polluted sites (such as the Grand Calumet River and the Indiana Harbor Ship Canal), reducing nonpoint pollution (such as fertilizer runoff), combating invasive species, restoring nearby habitats, and expanding wetlands in the Great Lakes basin. That same year President Barack Obama appointed Cameron Davis, president of the Alliance for the Great Lakes, as the first-ever Great Lakes "czar" to coordinate the various programs.[7]

Lake Michigan's water quality has greatly improved since the Clean Water Act was enacted. But it is not yet as good as it will be.

BEACH CLEANUP

Every September, the Alliance for the Great Lakes organizes a beach cleanup event in which thousands of volunteers remove trash from Great Lake beaches while collecting data on what they find. This annual event is part of the International Coastal Cleanup, which involves more than five hundred thousand volunteers in ninety-one countries.

Fishing in Lake Michigan, circa 1920. *Calumet Regional Archives.*

In September 2015, students from Willowcreek Middle School and Nativity of Our Savior School worked with undergraduates from Valparaiso University in a program called Building Bridges for Environmental Stewardship: Schools, University and Community Collectively Embracing the Health of a Local Watershed. Working together, the students participated in the international beach cleanup event along the Lake Michigan shoreline, and also conducted field studies at Imagination Glen and in Salt Creek.

LAKE MICHIGAN FISHING

Commercial fishing in the nineteenth century was big business in all three lakefront Indiana counties. Fishing at Michigan City started in the 1830s, in the Miller area before 1872, and at Waverly Beach (today Dunes State Park) beginning in 1907.[8] The fish caught—native whitefish, trout,

Facing: Willowcreek Middle School students working with VU students participate in the 2015 International Coastal Cleanup operation.

herring, and even sturgeon—were both sold locally and shipped by rail to Midwest cities. On occasion, the number of fish caught in a day was so great that customers couldn't be found for them all, so they were buried in the ground.

However, historian Jasper Packard wrote in 1876 that the best years for whitefish and trout fishing had been 1856 and 1857.[9] In 1904, Eugene Daniels noted a decrease in the fish population at Michigan City, and surmised that sewage dumped into the lake had driven the fish away. Or perhaps, because of overfishing, so many fish had been caught that the supply was simply exhausted.[10] Both factors took their toll.

However, another problem, unseen for many years, was the invasion of aliens! For thousands of years Niagara Falls prevented ocean-living fish from entering the upper Great Lakes. The Welland Canal bypassed the falls in 1829, giving both ships and fish a route from Lake Ontario to Lake Erie and ultimately to Lakes Huron and Michigan. Since pioneer times more than 180 nonnative species have come into the Great Lakes. Probably the greatest number came in ballast water in ships, which was emptied into the lakes as the ships were filled with lumber, rock, or other cargo. About a third have been brought in by people, often from aquarium water or bait buckets that have been dumped into the lake. Some of these "alien" species died out shortly, but others have survived, multiplied, and had detrimental effects on the native species.

The three worst alien species in Lake Michigan are probably the sea lamprey, the alewife, and the zebra mussel. Sea lampreys, which are native to the Atlantic Ocean, were first seen in Lake Michigan in 1936. They resemble eels, but they are a very separate species. Sea lampreys are fish that have a mouth that is a large sucking disk with sharp teeth surrounding a razor-sharp tongue. They feed on many species of fish, including trout, whitefish, salmon, and walleye, by attaching to the fish and sucking out its bodily fluids. Each adult lamprey can kill about forty pounds of Lake Michigan fish during its life.

Within thirteen years of that first sighting the lampreys were widespread, and as they multiplied, the number of lake trout dramatically declined. Like salmon, sea lampreys swim upstream in the lakes' tributaries to lay their eggs, and it is there that control efforts are being undertaken.

The mouth of a sea lamprey. *US Fish and Wildlife Service.*

A sea lamprey, *Petromyzon marinus.*

A Trail Creek coho caught and photographed by Don Calhoun.
Indiana Department of Natural Resources.

The primary method to control sea lampreys is through lampricides, which kill lamprey larvae before they develop their deadly mouths.[11] A newer nonchemical method is the use of stream barriers such as the one built on Trail Creek in 2012. These control methods have successfully reduced the sea lamprey population to a fraction of what it was before the controls were started.[12] Unfortunately, the sea lampreys have not been eliminated and thus the controls have to be maintained.

In 1949 the alewife, another Atlantic Ocean native, was unintentionally introduced into the lake. Alewives are voracious eaters and they ate the same food that many of the native fish ate, resulting in a number of local extinctions.[13] The alewives thrived until they literally ran out of food. By the mid-1960s, millions of stinking dead alewives were being washed up and deposited on Lake Michigan beaches. Not surprisingly, the stench caused a decrease in shoreline tourism.[14]

To combat the alewives, in 1966 and 1967 the state of Michigan introduced Pacific Northwest coho and chinook salmon into the lake. Since

The barrier across Trail Creek. The trap that bypasses the
barrier is on the left. *Northwest Indiana Steelheaders.*

then salmon and trout have been stocked along Indiana's shoreline from
Whiting to Michigan City, including Trail Creek and the east branch of
the Little Calumet River, by the Indiana Division of Fish and Wildlife.[15]
Although Michigan's efforts were controversial, the introduction was
successful from the fisheries' point of view. The salmon were aliens, but
then the alewives were also.

Zebra mussels, which arrived in the Great Lakes in 1986, are small
creatures that filter many of the microscopic organisms in the lake. Like
the alewife, these mussels consume the same food resources that our
native fish have depended on for thousands of years.

A Department of Natural Resources employee with a trapped lamprey.
Northwest Indiana Steelheaders.

Salmon jumping at the Trail Creek barrier. *Northwest Indiana Steelheaders.*

TRAIL CREEK

Trail Creek, located in and south of Michigan City, is considered by many to be one of the best trout and salmon fishing streams in the country. Several types of fish from the Salmonidae family can be found there, including skamania steelhead (a trout-salmon hybrid), rainbow, steelhead, and brown trout, and coho and king/chinook salmon. The creek and its two forks are long enough that they provide many opportunities for anglers' different preferences, including fly fishing, spin fishing, and drift/float fishing.[16] The Trail Creek watershed drains approximately fifty-nine square miles of LaPorte County.

The Indiana Division of Fish and Wildlife stocks chinook and steelhead every spring and coho in the fall.[17] Northwest Indiana salmon and trout live most of their lives in Lake Michigan, but when they mature, they find the stream where they were stocked or hatched. Most of the fish that return to Trail Creek and the Little Calumet River each year are chinook and coho salmon and steelhead trout. Along with sea lampreys, they travel upstream to spawn. According to the Indiana Department of Natural Resources (IDNR), Trail Creek unfortunately used to produce tens of thousands of sea lamprey larvae each year.[18]

In the spring of 2012 IDNR installed a sea lamprey barrier across Trail Creek at Springland Drive. In season, visitors to the area can see the salmon jumping over the barrier that blocks the lampreys. The lampreys and the weaker salmon swim into the trap at the near side of the barrier, where the lampreys are removed and the salmon released. The barrier saves the Great Lakes Fishery Commission the cost of chemical treatments. Not surprisingly, fishing in the area around the barrier is prohibited.[19]

Among the benefits of barriers such as the one on Trail Creek are reduced costs of purchasing and applying lampricides and simply a reduction of chemicals in the water that could adversely affect others in the environment. The disadvantage is that barriers interrupt stream flow for fish, invertebrates, canoers, and kayakers.

Brownfields Restored to Usefulness

12

BROWNFIELD REDEVELOPMENT

A brownfield is land that was previously used for commercial or (more likely) industrial purposes and is now the site of real or suspected contamination. Large brownfields may have been home to abandoned factories; a small brownfield may have once had a gas station. Many former brownfield sites remain vacant for decades, either because their locations are not desirable to new developers or because restoring them would cost more than the land would be worth after the work was done.

The reuse of brownfields has become both a national and a Calumet Area priority. Such reuse improves the environment and thus may prevent air and water contamination spread, eliminates a source of blight and therefore increases the value of neighboring properties, and helps prevent the unnecessary development of open spaces elsewhere.

In 1995 Congress established the EPA's Brownfields Program, which empowers communities and other stakeholders to collaborate in assessing, cleaning, and sustainably reusing brownfields. The program provides grant money for environmental training, assessment, and planning, and to capitalize loans in order to obtain funds to restore brownfields.[1] It also empowers communities to address brownfields and think creatively about possible ways to reuse land.

The Indiana Brownfields Program was established in 1997. It offers educational, financial, legal, and technical aid and works with the EPA and the various stakeholders to assist Indiana communities in making their brownfields productive again. To date, the program has supported 729 projects, which have helped create 298 new businesses, retained 243 other businesses, created 14,169 new jobs, and retained 5,353 other jobs.[2]

Most Calumet Area brownfields are located in industrial sections of the region's lakeshore cities. Several of the cities have established environmental or planning departments, which continue to wrestle with how to determine the best ways to deal with their abandoned industrial properties.

The Northwest Indiana Brownfields Coalition was formed specifically to assist the cities of Gary, Hammond, and East Chicago in finding good new uses for their brownfield sites. In recent years there have been a number of successes. The West Point Industrial Park property in Hammond was one of the first sites to receive assessment assistance from the Brownfields Program.[3]

MUNSTER STEEL COMPANY

In 1957 when Munster resident O. C. Robbins decided to establish the OCR Steel Company, he purchased a few acres from the National Brick Company and built his plant on the south side of town. The location was perfect—the south end of Calumet Avenue, away from residential neighborhoods, and between the Pennsylvania and Grand Trunk Rail-

Munster Steel Company. *Munster Historical Society.*

Munster Steel's new, efficient plant at Hammond's West Point Plaza Industrial Park.

road tracks. The location made delivery by both train and truck possible. In fact the Pennsylvania line (later the Penn Central) supplied material to the company for several decades. The plant's only neighbors were agricultural fields and the National Brick Company plant and clay pits, some of which were used for dumping trash—and most of those were on the other side of the Pennsylvania tracks. The company was later incorporated as the Munster Steel Company.[4]

Munster Steel is a fabricating facility, a plant that grew from a small one-building operation in 1958 to a 140,000-square-foot multiple-building facility by the 1990s, employing more than forty workers and fabricating more than eight hundred tons of beams, girders, and supports per month.[5] The company prepares steel for bridges, steel-framed buildings, and other structures. Among its more glamorous products are the steel in Trump Tower; the Chicago Transit Authority's blue and brown lines; skyboxes for the White Sox, Cubs, and Bears; the Wells Street renovation over the Chicago River; and University Center at Portage, Indiana.[6]

By the 1990s, however, the plant's neighborhood had changed. The Pennsylvania tracks were gone and a bike trail was planned for its route. The nearby fields now sprouted houses and shade trees. The brickyard was gone and was soon replaced by a park shelter and soccer field, and some of the abandoned clay pits had been converted into the town's sanitary landfill. In the next decade the landfill would be closed and the grounds converted into a nine-hole public golf course with a new clubhouse featuring a view of Maynard Lake, occupying one of the former clay pits. Calumet Avenue had been widened, and then in 1999 work finally began on the extension of the busy street to US 30 in Dyer. When that was completed in 2004, traffic on the street grew much heavier. The steel company property had become land desired by commercial developers.

After several years of discussion and negotiations, the family-owned business, then in its third generation of leadership, decided to move. After extensive searches for possible new locations, including a cornfield west of Lowell, the company decided to relocate to Hammond's West Point Plaza Industrial Park and purchased twelve acres of land there.[7] The land had already been cleaned up, and the city helped plan the company's brand new facility. The big move was made in 2014. Now in Hammond, Munster Steel has an efficient and environmentally sensitive building. Waste steel shavings are collected and recycled. Painting is done in a separate room where fumes are pulled through filters and can't escape into the outside air. Steel rails used to bring in new work are embedded into the concrete floor so that they don't create a tripping hazard. All rainwater is diverted into two retention basins and does not enter the Grand Calumet River or city sewers. Munster Steel's decision to move to a brownfield site, and its creation of a new plant that does not foul its new community's air or water, is a model for other companies to follow.

LOST MARSH GOLF COURSE

The city of Hammond initiated the Lost Marsh project in 1997 when it acquired a one-hundred-acre dumpsite east of Wolf Lake and south of George Lake. The area was found to contain a large amount of industrial fly ash and about 3.5 million cubic yards of slag from local steel mills.[8] Af-

Lost Marsh Golf Course, with George Lake in the background.

The Lost Marsh clubhouse.

ter essential environmental assessments were done, the city capped the former slag dump using biosolids that it obtained from the Hammond Sanitary District and sand dredged from the South Basin of George Lake (which the city already owned).

The site now features an eighteen-hole golf course with a clubhouse/ restaurant building with meeting rooms that are available to community groups. The golf course and the driving range both provide revenue. A nine-hole youth golf course also built on the property is used for the First Tee program, an international youth program that teaches golf and life skills to students from area schools. An empty house was moved across the site to its current location and turned into a youth clubhouse that now contains lockers as well as an indoor driving and putting range. All this is next to an outdoor driving range that is available to both young people and adults.[9]

Besides the two golf courses, the Lost Marsh development has an area set aside for bird watching and a public fishing pier in the South Basin of George Lake. In addition, through the late fall and winter the golf cart trails are also available for biking and hiking. And when weather and snow conditions allow it, the golf course hills are used for sledding. The city now also owns the North Basin of George Lake, which is being kept in its natural state as was suggested by citizens during numerous public hearings.[10]

The Lost Marsh project received the Indiana Governor's Award for Environmental Excellence in 2002. Because of the city of Hammond's reuse of biosolids, it was also featured at the World Environmental Summit on Sustainable Development held in 2003 in Johannesburg, South Africa.[11]

The Lost Marsh clubhouse and restaurant, which sits at the top of a small rise, was designed in a distinctive Frank Lloyd Wright style. The restaurant, which includes some open-air seating, and the bar are open daily to the public. The building also has a banquet hall, which is rented for special events.

AN UGLY BRICKYARD AND LANDFILL
BECOME A DESTINATION PARK

Stage One: Brickyard and Strip Mine

About fourteen thousand years ago, when Lake Michigan was first formed, what is now Munster lay under about twenty feet of water. Too far from the shore to have sand deposited in what is now central Munster, mud settled down to the bottom of the lake and formed a thick layer of so-called "blue rubber" clay that is still there. When the lake level dropped to where Ridge Road is now and a sandy beach developed there, the area south of the ridge remained a rather flat wetland.[12]

The Cincinnati and Chicago Air-Line Railroad (later the Pennsylvania) built its line through the area in 1865. In 1880 the Grand Trunk line

Employees at Munster's National Brick Company. *Munster Historical Society.*

The brickyard area at "Maynard," 1929. The red north–south highway, Ade Way, or State Highway 10 (later US Route 141), is today Calumet Avenue north of the brickyard and Columbia Avenue south of the bend in the road. *US Geological Survey Quadrangle,* Calumet City, Illinois.

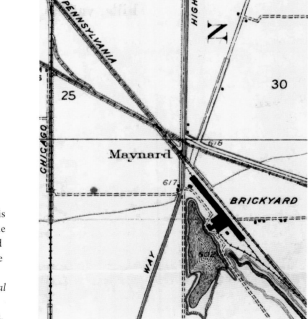

The abandoned (or about to close) American Brick
Company plant, circa 1987. *Russell Snyder.*

was built, crossing the Pennsylvania just west of where Calumet Avenue
is now. That same year Aaron Hart drained the swampy land by having
the nearby ditch, which was named for him, dug.

Then in 1906, taking advantage of that rich and nearly pure clay subsoil, the National Brick Company purchased 160 acres of land, erected a
brickworks southwest of the Pennsylvania line, and removed the topsoil,
sending as many as eight hundred carloads of it to Gary, which needed it
for its parks and residential neighborhoods. Clay was dug from both sides
of the tracks, hauled to the plant, where it was made into bricks, and then
shipped to waiting customers, many of them in Chicago.

By 1939 the National Brick Company had the capacity to make three
hundred thousand bricks a day.[13] The downside was the creation of a
huge strip-mined area. As the pits got deeper, National Brick had to use
pumps to keep them dry. After 1930, when any one pit was mined out,
the company would start filling it with trash. For many years, the town
of Munster had permission to dump and burn its garbage there. After
1958, Dyer's and Lansing's trash was accepted as well.[14]

In 1966 the company was sold and renamed the American Brick Company. By this time houses were being built nearby and residents were becoming upset about the smoky plant. Of course, the plant was there first, but its neighbors complained loudly.

In the years following the passage of the Clean Air Act in 1970, the company spent more than $1 million on new kilns, filters, and scrubbers, but it couldn't solve its pollution problems. It ceased operations in 1987 and sold the land to the Community Foundation, which deeded it to the town two years later.[15]

Stage Two: Landfill

In 1968, the rather new Munster Park Department purchased 141 acres of the old strip mine and dump with the intention of eventually turning that property into a park. However, the intermediate use would continue to be as a landfill. At a time when open dumping and burning of garbage was still common in the Midwest, the Munster Landfill became a model of how landfills should be processed.

Although not quite a park yet, winter sledding began in 1979 on "Mount Trashmore," a hill composed of buried garbage towering some eighty feet above Calumet Avenue. Unfortunately the still-decomposing garbage underneath tended to warm the surface and melt the snow faster than area children wanted. In summer, the hill became a part of mountain bike competitions.[16]

The landfill accepted garbage from Munster and nearby communities, and its operations earned a nice income for the town. But with more subdivisions being platted right next to the landfill, residents called for it to close.

Stage Three: Destination Park with Green Attributes

The last section was closed in 2004, and after receiving public input, the Park Board had a master plan created for what would be the two-hundred-acre Centennial Park. The new park, which doubled the amount of park space in town, features a nine-hole golf course and driving range over the landfill, two miles of walking trails, a dog park with fountains and play equipment, an entertainment stage, a sculpture garden, formal

Native plants surround Maynard Lake, a former clay pit, and
the LEED-certified clubhouse at Centennial Park.

gardens, and a clubhouse. The large waterlogged clay pit was reshaped
into the forty-acre Maynard Lake, with its three ponds and two bridges,
named for the area's historic name. Two shelters and two soccer fields
were built where the old brickworks had been located. And although not
quite finished, it was dedicated on July 1, 2007, one hundred years to the
day after the town's 1907 incorporation.

The clubhouse is LEED (Leadership in Energy and Environmental
Design) certified by the U.S. Green Building Council at the Silver level,
meaning that more than 50 percent of construction materials were ex-
tracted regionally, more than 5 percent of construction materials came
from recycled content, 85 percent of its construction waste was diverted
from landfills, and a green roof was installed that reduces storm water
runoff and cooling demands for the building. The clubhouse has a pro
shop and a banquet hall and restaurant, both overlooking Maynard Lake.
Centennial Park sits astride the Pennsy Greenway, a bistate walking
and bicycle path that follows the old Panhandle/Pennsylvania Railroad
route from Lansing, Illinois, to Crown Point.

Jake and Heather Schoon's wedding party hams it up on one of the two
bridges over Maynard Lake at Centennial Park. Once a strip mine and
probably the ugliest area in Lake County, the park is now a destination
site for weddings and other celebrations. *Jake and Heather Schoon.*

When developed, Centennial Park was the only park in Indiana to be
formed upon a capped landfill. In 2007, the town of Munster was given
the Indiana Association of Cities and Towns' annual Award for Community Achievement for its development of Centennial Park.[17]

Finally, the park has an energy generation station, made possible with
a $1.9 million grant from the US Department of Energy, whereby meth-
ane from the landfill is used to generate electricity. Its Jenbacher engine/
generator produces about 1.1 megawatts of electricity. As a result the
landfill's carbon footprint has been reduced by 5,500 metric tons of CO_2
annually, which is the equivalent of removing from the air the emissions
from 1,100 cars, generating power for 720 homes, planting 1,200 acres of
trees, or recycling 2,000 tons of waste. In addition, the system has a heat

The Jenbacher engine/generator at Centennial Park.

conversion component, which takes excess heat from the generator and transforms it into additional electricity, which is then sold to NIPSCO.[18]

HOBART BRICKYARD BECOMES HIDDEN LAKE

Hobart, once one of two centers of brickmaking in the Calumet Area, was the first industrial city in Lake County. It had several brickworks, all of which mined the rich and pure glacial clay that was found in a forty-foot-thick deposit beneath the city, but close to the surface in the area just north of George Lake. By the late 1860s Hobart had four brickyards in operation. The two largest yards from the late 1880s through the 1910s were the Kulage Brickyard south of the Pennsylvania tracks and the Owen Terra Cotta Works north of the tracks.

Local historian George Garard wrote in 1882 that "Hobart owes its prosperity more to its brick yards than to anything else."[19] When the National Fire Proofing Company (NATCO) closed its plant, this site,

The Hidden Lake subdivision and park are located on a former brickworks site with a fairly deep old clay pit.

which was located very close to the historic downtown area, was not a beautiful one. But today it's quite different. A subdivision with single-family homes, townhomes, condominiums, Hidden Lake, and a park with a swimming pool and picnic shelter occupies the space.

PORTAGE LAKEFRONT AND RIVERWALK

The new parkland next to Burns Waterway is a stunning example of how a polluted industrial wasteland can be converted into a clean recreational area. The restoration of this area was done through partnerships and a coordination of efforts by the National Park Service, the National and U.S. Steel companies, and the city of Portage. The park's land and structures are owned by the National Park Service but are managed and operated by the city of Portage.

Portage Lakefront and Riverwalk (left of Burns Ditch)
before restoration. *National Park Service.*

Back in the early twentieth century, much of the area was sand mined
by the Consumers Company. Then in 1926 Burns Ditch was excavated,
providing a direct route to Lake Michigan for the waters in the east
branch of the Little Calumet River. The area along the beach was subdi-
vided and several summer cottages were built in the area east of Ogden
Dunes. Several organizations, including the Gary Boat Club, purchased
rights to use the shore of Burns Ditch to tie up boats. The club also built
a clubhouse on the beach.

Portage Lakefront and Riverwalk after restoration.

Land on both sides of Burns Ditch was purchased by the National Steel Company as early as 1929. In 1959 it built a mill on the east side of the ditch, using the west side for its wastewater treatment facility and for its open artificial lagoons where industrial acids and other liquids were dumped.

In 1986, Congress included this fifty-seven-acre property in the expansion of the Indiana Dunes National Lakeshore. However, before the National Park Service could purchase the land, it had to be cleaned up. Working with the EPA, the National Steel Company initiated the removal of all the hazardous wastes and the restoration of the area. The company agreed to go beyond legal requirements to meet cleanup standards for the planned public use of the area, a process that included removal of hazardous materials from two settling lagoons.[20]

In 2003 National Steel sold its mills, including this parcel, to U.S. Steel, which continued the cleanup process. Later that year the EPA and IDEM gave the site a "clean closure" designation, and the next year, no longer toxic, it was purchased by the National Park Service. With the site remediated, the natural forces of plant succession replaced the forced sterility of the toxic dump. The former settling ponds are returning to a high-quality habitat with a diverse assemblage of native plants.[21]

The National Park Service and the city of Portage together designed the site, which includes a pavilion, a fishing pier and breakwater walkway, trails and a riverwalk, and a 125-car parking lot. The city secured a grant of $10 million for the project from the Northwest Indiana Redevelopment Authority (RDA). The Development Advisory Board of the National Park Service praised the plan as a model of both ingenuity and cooperation. Congress reviewed and approved the plans in October 2007 and work began.[22]

The development plan was environmentally sensitive. The pavilion, trails, roads, and parking lot were located in areas that previously had been disturbed. For example, the new entrance road followed much of the route of a preexisting gravel road, and the parking lot is where the storage and staging areas had been. The drainage system was altered to provide a natural return of storm water to the ground through catch basins, curbs, rain gardens, and drywells rather than directly into either Lake Michigan or Burns Waterway.[23]

At the formal dedication ceremony in November of 2008, Congressman Peter Visclosky described the new park site as the linchpin of his Marquette Plan—that multiyear proposal to increase public accessibility to 75 percent of the Lake Michigan shoreline in Indiana. Soon 150,000 visitors a year were visiting the new Portage Lakefront and Riverwalk.

The state of Indiana contributed to the project by constructing an $11 million bridge over rail lines and US Route 12. The bridge provides unimpeded access to both the mills and the lakefront and riverwalk.

The Pavilion

At the top of a dune bluff overlooking Lake Michigan and Burns Waterway is the strikingly beautiful pavilion, which has a large activity room, a snack bar and patio, restrooms, and a visitor information desk. It has a unique design in which the curved roof echoes the curves of nearby dunes and waves, a soffit brings to mind a boat's hull, and a fireplace evokes hearth and home.[24] The pavilion was built using LEED standards and was awarded Gold-level certification, as it used locally produced

The pavilion.

recycled steel and other nontoxic materials, acquired the majority of materials for the site from less than five hundred miles away, and included reflective roofing, water and energy efficiency, and a glass frit pattern visible to migratory birds.[25]

The pavilion, which is located where the previous wastewater treatment plant stood, utilizes geothermal heating, low-flow water faucets, and motion-detector-operated lighting. The landscaping naturally included only native plants. In 2009, the park was named Indiana's "Outstanding Park Development" by the Indiana Parks and Recreation Association.

Within just a few years of its opening, however, the Portage Lakefront became so popular that visitors couldn't find places to park. The 125 available parking spaces turned out to be inadequate, especially on hot summer weekends, and so the city of Portage made plans for expansion.[26] A sixty-nine-acre plat south of the Portage Lakefront's southern boundary was purchased from U.S. Steel in 2013 and is being used for an additional parking lot that will have room for three hundred cars. A large steel company warehouse was demolished and, in partnership with the U.S. Army Corps of Engineers, was replaced by a low sand hill. Plans for upcoming years include the planting of native plants and the creation of a wetland area.[27]

SUPERFUND SITES

Superfund is the EPA program set up to handle the worst US hazardous waste sites. The name is also used for the fund that Congress established to help clean up such sites. The Comprehensive Environmental Response, Compensation, and Liability Act (CERCLA) was passed by Congress in 1980 in order to clean up sites contaminated with hazardous substances. It was amended in 1986 by the Superfund Amendments and Reauthorization Act, whose various changes included the addition of Section 121, specifying minimum requirements for cleanup operations. Under these acts, the EPA is required to ensure that hazardous waste sites are cleaned up. The EPA may handle the cleanup itself, or it may "compel responsible parties to perform cleanups or reimburse the government for EPA-led cleanups."[28]

With all the industry that has at one time or another been located in the Calumet Area, and because many of these corporations have gone out of business and many of those had neither the funds nor, perhaps, the inclination to clean up their properties before they left, it is not surprising that there are dozens of properties in the area contaminated with hazardous waste. It should be remembered that for approximately the first hundred years of industrialization, there were few or no regulations concerning pollution of the air, water, or land.

After assessment, the EPA puts the most hazardous sites on the National Priorities List. Superfund sites tend to be rather small parcels of land in industrial areas. One site, however, has affected a large residential area: in 2014, the Department of Justice announced a $26 million settlement between the EPA, IDEM, and Atlantic Richfield (ARCO) and DuPont for the cleanup of lead and arsenic contamination caused by the U.S. Smelter and Lead Refinery in the Calumet neighborhood of East Chicago. About three hundred residential yards and public properties are contaminated with lead and arsenic. The cleanup involves removing up to two feet of soil from many of the affected residential yards and replacing it with clean soil.[29] Fortunately, though, most Northwest Indiana Superfund sites are not on the National Priorities List.

Solid Waste and Recycling

13

RECYCLING IN THE CALUMET AREA

Residential recycling began slowly as a voluntary activity. During World War II, the whole country was recycling. Even before the war, a nation-wide search for aluminum had begun. On July 29, 1941, seven hundred children participated in a "pan swim" at North Township's Wicker Park Pool. Admission was an aluminum pot or pan, which then were melted down for the defense industry.[1] Women collected old leather items to be made into jackets for British pilots. Then, during the war, rubber, rags, metal, newspapers, magazines, cardboard, and even fat were all collected.

Recycling continued until the end of the war, with whole communities participating. In the then-little town of Munster alone, the last recycling drive of the year netted more than seven tons of goods.[2] After the war, recycling efforts slowed down and nearly came to a stop. But some companies, such as the Keyes Fibre Company in Hammond, continued to accept and use old newspapers to make new products. In fact 33 percent of Keyes's fiber usage came from cartons and wastepaper.[3] To earn spending money, many youth groups would collect used newspapers and take them to Keyes's plant on Indianapolis Boulevard.

Soft drinks in those days came in glass bottles and folks had to pay a deposit of two cents per bottle, which they got back when the bottle was returned to the store. But otherwise, the United States was becoming a throwaway society. After the first several Earth Days in the 1970s, however, some environmentally focused men and women became concerned about the problem of solid waste disposal and began to propose recycling as a way of reducing the amount of waste buried in landfills.

REGION RECYCLING AND CURBSIDE PICKUP

In 1974, a small group of people formed the Munster Recycling Committee and set up a recycling center at the former Nike site in town. The next year it expanded its focus and became Region Recycling, and in 1977 it became a tax-exempt nonprofit organization. With a cadre of volunteer helpers, the Region Recycling Center collected metals, paper, and glass weekly.

The organization's founder and president, Walter Helminski, started writing a weekly column in the *Times* urging people to recycle. Recycling in those first several decades was very time-intensive. On Saturday mornings local residents would bring their recyclables to the center on Columbia Avenue, where they would physically separate the items: newspapers into one dumpster, aluminum in a second, clear glass in a third, and colored glass in another. Plastics weren't yet accepted. Youth groups earned a little money for helping residents unload their cars at the center and dump each item in the proper bin.

In June of 1990, Munster became the first Calumet Area community to establish a town-wide curbside recycling program; it was also the first area recycling program to include some plastics.[4] In December of that year, the Metro Recycling Buy-Back Center was opened on town property near the entrance to the town's Calumet Avenue landfill. Months before the program began, the town launched an education and promotion blitz that included several mailings and a newsletter. Town officials visited schools and civic groups to explain the upcoming program. Town engineer Jim Mandon estimated that in total they talked to about 5,300 residents at these gatherings. The town anticipated that perhaps half its families would participate, but in the first week the level of participation was an astonishing 97 percent.[5]

Recyclables still had to be separated by the truck drivers. Crews would hang each recycling bin on the side of the collection truck, separate the materials, and scan the address label before placing the bin back on the ground. Between that process and the large number of residents participating, the actual collections that first week took much longer than predicted. But the recycling program successfully diverted about 15 percent of the town's solid waste away from the landfill. When figures from its composting program were also included, the amount diverted reached 30 to 32 percent.[6] The next year, Munster was given the Indiana Governor's Recycling Award.[7] (Years later, after the town privatized trash collection, all recyclables were placed into a single-bin truck at the curb, dumped commingled, and later separated.)

The year 1991 was big for recycling area-wide; the *Times* called it a "bandwagon" and a craze that was "sweeping the nation."[8] By the end of the year more than a dozen area communities had at least drop-off facilities that required residents to haul their recyclables to their centers and deposit items into the appropriate bins.[9] Hobart had citywide leaf collection and two drop-off bins. East Chicago had a curbside pickup program for a few neighborhoods and seven drop-off sites. Hammond had curbside pickup for about nine thousand households.[10] In addition to the city- and town-originated programs, the Bethlehem, LTV, and U.S. Steel companies collected scrap steel. And recycling companies, such as Metro, Indiana Recycling Services, and Worldwide OFC Recycling, had drop-off arrangements with several local governments.

By November of 1991 there were fifty-one recycling drop-off centers in Lake County alone.[11] Porter County had nine. Some centers were publicly run; others were privately run. Articles about recycling made the front pages of local newspapers.

SOLID WASTE MANAGEMENT DISTRICTS

County-wide (or multicounty-wide) new units of local government called solid waste districts were mandated by Indiana House bill 1240, which was signed into law in March of 1990. While counties were permitted to either form their own district or join with other counties in a single district, most counties in Indiana, including all three in the Calumet Area, chose to form single-county districts. The Lake, Porter, and La-Porte County districts were established in 1991; their formal operations began in 1993. Although Lake County had recycling centers across the county, most were drop-off programs and they were diverting only 17 percent of the county's waste from landfills. Porter County's nine centers plus curbside recycling in four communities resulted in diverting just 5 percent of its waste.

Each district was to coordinate its solid waste disposal and to develop a twenty-year plan whereby recycling programs would reduce the amount of waste going into landfills by 35 percent by 1996 and 50 percent by 2001. A formula was established to calculate this percentage based on baseline waste generation and included figures for recycling, composting, source reduction, reuse, and economic and population change. To meet these goals, all three districts established extensive public information and education programs. Indiana did not reach its state-designated goals. The state achieved only a 30 percent reduction by 1996 and a 39 percent reduction by 2001. Other states' efforts weren't much better. In fact a recycling rate of 40 percent was found to be quite difficult to attain.[12]

A state program review in 2003 found that while it may appear that disposing of waste in landfills costs less than recycling, recycling still had benefits in addition to financial ones: it delayed the filling of landfills, reduced the use of virgin materials, provided a safer method of disposal for some materials, and reduced greenhouse gas emissions. The review's conclusion stated that Indiana's overall waste disposal costs would have been higher without recycling.[13]

Due to concerns about inaccuracy surrounding the original baseline waste-generation figures, the waste diversion goals of 35 percent and 50 percent were dropped in 2012 and replaced in 2015 with a new 50 percent recycling goal coupled with new data-reporting requirements.

Calculating the amount of material recycled and landfilled and computing a percentage of each (per the 1990 regulation) is often difficult and never completely accurate. One reason for this is that many communities employ private contractors for these services, and those services may or may not be able to calculate how much waste was collected from various municipalities. Second, some private organizations collect recyclables (such as aluminum and paper) and sell the tonnage collected. This material, repurposed materials, and organic material composted on-site are not figured into the amount of material diverted from landfills. Nevertheless, some communities can and do keep records of garbage, composting, and recycling tonnage collected within their boundaries. Two that

Grinding up branches for compost. *LaPorte County Solid Waste District.*

do in Porter County are Portage and Valparaiso, which in 2014 reported an excellent 56.4 percent and 40.4 percent diversion rate, respectively.[14]

All three counties have recycling drop-off stations for various products such as electronics, tires, and appliances. Unneeded medications are collected at several law enforcement stations. All three counties also have programs for composting yard waste—some have curbside programs, and others have drop-off composting centers. The three counties collaborate with their joint household hazardous waste programs.

All three districts have developed free education programs for schools (which are discussed in greater detail in chapter 14). In addition, all three have also developed some very interesting and innovative other programs. Lake County, for instance, has a "ReUZ Room" in its environmental education center in Hammond where residents (primarily teachers) can come and help themselves to donated craft materials.[15] In Hobart, latex paint can be dropped off and the mixed paint purchased for three dollars per gallon.[16]

Porter County, meanwhile, lends and picks up special recycling bins for special events. It also has an "Adopt a Road" cleanup program, sells rain barrels, and collaborates with Hobart on its latex paint program. And LaPorte County offers curbside recycling to all county residents.

The Hobart paint recycling unit.

Twice a year there is also a "five-in-one" collection at the LaPorte County Fairgrounds, which accepts five different types of recyclables such as tires and appliances. The county also features on its website a very detailed list of what can and cannot be included in curbside pickup.[17]

DIVERSIFIED RECYCLING

The recycling process is much different today than it was in 1990. Then, residents had to separate their recyclables first and each town was on its own to find customers for its collections. Today, most of the recycled materials picked up in Lake, Porter, and LaPorte Counties are shipped to the Diversified Recycling plant in Homewood, Illinois. This privately owned company is one of the largest of its kind in the Midwest.

This company's single-stream system allows it to accept recyclables such as junk mail, magazines, cardboard, newspapers, tin and aluminum cans, and glass and plastic bottles and jars, all mixed together. At the plant, the material is sorted through a series of screens and sorting machines, as well as by its employees. (This is why residents themselves are no longer required to separate materials before they are collected.) The sorted recyclables are then shipped to various domestic and international markets. Since 2000, Diversified Recycling's Homewood facil-

The Diversified Recycling plant in Homewood, Illinois. *Diversified Recycling.*

ity has processed more than 1,100,000 tons of recyclables. From April to September, the facility offers free tours for school groups of fifteen students or fewer, ages sixth grade and above.[18]

NONGOVERNMENTAL RECYCLING OPPORTUNITIES

Nongovernmental for-profit and not-for-profit recycling opportunities are found throughout Northwest Indiana where batteries, eyeglasses, CFL light bulbs, plastic bags, or other materials are collected. For example, Green Tree Plastics of Evansville, Indiana, is a private company that turns bottle caps into park benches made entirely of recycled plastic. Various nonprofits in the area collect the caps and transport them to Evansville for the transformation. And reuse has always been alive and well in Northwest Indiana, in the sense that the area has for decades had plenty of antique stores, secondhand or resale shops, garage and yard sales, nonprofit rummage sales, and used book sales.

Plastic lids and bottle caps, which, when combined with four hundred pounds of others, will become another park bench like this one.

LANDFILLS

As of this writing there are no still-operating final solid-waste disposal facilities in any of the three Northwest Indiana counties except for one construction and demolition landfill in Lowell. The last operating sanitary landfill was the one in Munster that closed in 2004. There were, however, transfer stations where collected garbage is put into large semis for shipping to more distant locations. Garbage is then sent to landfills in Indiana, Illinois, and Michigan.[19]

Local Pioneering Environmental Heroes and Heroines

14

HENRY CHANDLER COWLES:
THE FATHER OF NORTH AMERICAN ECOLOGY

Henry Chandler Cowles was born on a Connecticut farm in 1869. It is said that he developed his fascination with plants and trees during his childhood there. By the age of sixteen, he was well acquainted with the plants in the area and had read Gray's *Manual of Botany*. After earning his bachelor's degree in 1894 at Oberlin College in Ohio, he went to Chicago to study geology.

Once at the University of Chicago, he realized that his true calling was really in the biological sciences, particularly botany and the new science of ecology, the study of the relationships between organisms and their environment. At that time ecology was a little-known discipline centered in Germany and Denmark—with no English-language texts yet available.[1]

Henry Cowles visited the Indiana Dunes for the first time in April of 1896 and spent much of the next two years learning more about the fascinating Duneland plant life. His PhD dissertation on the plants of the Dunes was the first major study of ecology in North America. Working in an area where the Glenwood-, Calumet-, and Tolleston-era dunes were so close to one another, he could readily see how the flora of the older dunes was different from that of the younger dunes. He discovered the effects of time on plant life and provided convincing evidence for the theory of succession.

Cowles's entire career was spent with the university's Botany Department.[2] He helped start the Prairie Club in 1911, was a member of

Henry Chandler Cowles in the Indiana Dunes,
1913. *Calumet Regional Archives.*

the Friends of Our Native Landscape founded by landscape architect
Jens Jensen, organized the Wild Flower Preservation Society, and was a
member of the Chicago Geographical Society and served as its president.
These organizations all actively supported the Illinois state park system,

Members of the International Phytogeographic Exchange rest during their tour of the Indiana Dunes. *National Park Service.*

the Cook County Forest Preserves, and the National Dunes Park Association. Northwest Indiana was Henry Cowles's laboratory.

Cowles spent much of his time in the Dunes and became very familiar with many parts of the landscape. Apparently he was fascinated by the sometimes "quaking" wetland west of Mineral Springs Road where the thick vegetation floats on water. After he got his doctorate and was admitted to the faculty at the university, each year he took a new group of students to see it. In 1913, when a group of European scholars let it be known that they wanted to see the Dunes, he took them to see this rare ecosystem as well. Henry Cowles made this wetland famous.[3] So, since it didn't have any other name, by the 1920s people were already referring to this feature as "Cowles Tamarack Swamp" and "Cowles Bog"—this in spite of the fact that this wetland actually is not a bog at all. Rather, because it has an influx of lime-rich water from springs beneath the floating mat, it is a fen.

Just sixteen months after the Save the Dunes Council was formed, it took advantage of a tax sale and purchased fifty-six acres of that rather famous wetland.[4] Eleven years later, when the US Department of the Interior announced that it would accept nominations for a register of environmentally unique parcels of land to be called "natural landmarks" (similar to the National Register of Historic Places), the council nominated what it called "Cowles Bog." The nomination was approved in 1965. Cowles Bog and LaPorte County's Pinhook Bog were the first "bogs" to be designated as landmarks. And with this official designation, the Cowles Bog name became official as well.[5]

Cowles loved taking students and fellow scholars to the Dunes and advocated for the creation of a "Sand Dunes National Park."[6] In 1920, he was one of the organizers of the first National Park Conference, whose goal was to encourage all the states to make the purchase of land for parks a high priority.

He mentored many University of Chicago students, including Victor Shelford, who did a study of tiger beetles in the Indiana Dunes. In 1946, as an ecologist at the University of Illinois, Shelford initiated the founding of the Ecologists Union, an organization committed to saving threatened natural areas. Four years later the organization's name was changed to The Nature Conservancy. It is now the world's leading conservation organization.

When Cowles retired from the university in 1934 he was the chairman of its Botany Department. He died at his home near the university in 1939. In addition to the Cowles Bog, the main lodge at the Dunes Learning Center in Porter, Indiana, is named for him.

FRANK V. DUDLEY:
PAINTER AND PROMOTER OF THE DUNES

Before the advent of color photography, the only way to capture the full beauty of the Dunes was in painting. And no one did a better job of doing this than Frank Dudley.

When Frank was nineteen he moved to Chicago and enrolled in classes at the Art Institute.[7] He became an active member of the Society of Western Artists, an organization that also included famous sculptor Lorado Taft, cousin of President William Howard Taft and future father-

Frank Dudley (1868–1957) at the Prairie Club fountain at Indiana Dunes State Park.
Photo by Arthur Anderson, held at the Calumet Regional Archives.

in-law of Senator Paul Douglas. Dudley exhibited at the society's shows and became better known and respected as an artist. His works were regularly shown at the Art Institute, and he was also a charter member of the Cliff Dwellers Club, whose other members included Frank Lloyd Wright and Jens Jensen.[8]

Unable to make his living as an artist, in 1908 he opened up the Dudley Shop on East Sixty-Third Street, where he sold cameras and art supplies. Beginning in 1910 he started doing some paintings of the Indiana Dunes, and once he saw the Dunes in autumn, he knew that he had "found his theme."[9]

Frank Dudley tagged along on several Prairie Club hikes, but never joined that organization; instead he joined the Friends of Our Native Landscape. He did support the National Dunes Park Association and attended the association's 1917 pageant in the Dunes. In May of that year he showed thirty paintings in a one-man show, *The Sand Dunes of Indiana and Vicinity*. In 1919, *National Geographic* magazine reproduced four of his paintings in its article about the Dunes. In 1921 Dudley built a cottage, Duneland Studio, near the base of Mount Holden, and he and his wife started having open houses on Sundays when the public could visit and purchase his work. In 1922, he, Bess Sheehan, and Henry Chandler Cowles together hosted about a dozen state representatives and senators at the cottage—Dudley used more than forty of his paintings to describe the Dunes and Cowles discussed the scientific and cultural value of creating a state park there. In 1923 the park was authorized and in 1925 the first land was purchased.

The Dudley studio was within the boundaries of the state park that he had promoted so vigorously. When the state then started purchasing property for the park, the Dudleys were invited to remain on park property, paying an annual rent of one framed painting. Thus, for the last several years, they were the only residents within the restricted area of the park. The Dudleys would come out to the Dunes in the spring and sometimes stay until December.

From 1927 through 1956, Dudley showed up to forty-five paintings at each of his fifteen one-man shows, with nearly every painting being of the Dunes area. In 1950 five hundred people attended his opening and he sold twenty-five paintings at that one show.[10]

Frank V. Dudley, *Sandland's Even Song*, 1920. Oil on board, 38 x 50 in. Sloan Fund Purchase, Brauer Museum of Art, 2001.44, Valparaiso University.

Throughout these years, Frank Dudley remained an active environmentalist. In 1954 he joined the advisory board of the Save the Dunes Council. He remained a member of the Friends of Our Native Landscape for the rest of his life, even serving as a director of the Illinois chapter from 1955 to 1957. After a three-day illness, Frank Dudley died at his Chicago home on March 5, 1957.

Five years after his death, many of the sixty-nine members of the Save the Dunes Council who traveled to Washington to testify for the creation of a Dunes national park took with them several of Dudley's paintings, which they placed around the perimeter of the hearing room. Dudley's paintings spoke for him. They still do.

BESS SHEEHAN:
MOTHER OF INDIANA DUNES STATE PARK

Without the efforts of Bess Sheehan, it is doubtful that Indiana Dunes State Park would ever have been created. Time was of the essence: Duneland property was being developed at a rapid rate and the price of the land was escalating. Bess Sheehan knew what to do and she did it.

Bess Vrooman Sheehan was born on April 16, 1882, in Jackson, Michigan. She attended the University of Michigan where she earned both a bachelor's degree in 1904 and a master's degree in 1905. She taught history in Big Rapids, Michigan, for a couple of years, and then in 1908 she moved to Gary to teach history in the already nationally recognized progressive Gary High School. She taught for four years until 1912, when she married Frank J. Sheehan, an attorney, who for a while also served as the judge of the Lake County Juvenile Court.

To merely say that Bess Sheehan was a civic leader doesn't come close to indicating the extent of her efforts and influence. Locally, she served as the chair of the Campbell Settlement House board of directors,[11] the campaign chair for the Gary YWCA, and the chair of the Red Cross Annual Membership Roll Call. Her greatest impact, however, came from her service with Indiana's Women's Clubs and the National Dunes Park Association.

In 1916, when the Prairie Club decided to form the National Dunes Park Association to advocate for a national park in the Dunes, Bess Sheehan was elected its secretary. The next year, she helped plan the mammoth pageant, *The Dunes under Four Flags*, as well as newsreels about the Dunes that were shown nationwide in movie theaters. Then in 1919, when the organization foundered, she turned her attention to creating at least a state park in the Dunes. Just two years after women in the United States were granted the right to vote, the Indiana Federation of Women's Clubs, which represented six hundred chapters in the state, had become a powerful lobbying group. Bess Sheehan was the state president.

Sheehan personally lobbied state legislators, and on January 26, 1923, before a special joint evening session of the General Assembly, she gave a

Right: Bess Sheehan (1882–1968).
Cannon, Loring, and Robb,
History of the Lake and Calumet
Region, *opposite 680.*

Below: Bess's reward: kids and adults
having fun at Indiana Dunes State
Park.*Calumet Regional Archives.*

two-hour stereopticon lecture to the legislators and many of their wives. Her presentation eloquently demonstrated both the beauty of the area and the arguments for its preservation—noting that the stretch of dunes available for inclusion in the park had already shrunk from fifteen miles to just six.[12] Her efforts finally paid off when, several months later, on the last day of the session, the Assembly approved the park.[13]

Her civic work didn't end there. Bess Sheehan later served as chair or director of several national movements, including three years as chair of the Natural Scenery Committee. She was appointed by Indiana's governor to the State Library and Historical Board in 1925 and served until 1933, and she helped plan the State Library building in Indianapolis. In addition to all of this, Bess Sheehan also authored a couple of booklets, including *The Northern Boundary Line of Indiana* in 1928.

Bess Sheehan was called the "Dunes Lady," but truly she is the mother of Dunes State Park. The Indiana Federation of Women's Clubs called her the best-known woman in the state, and the Chicago Geographical Society made her an honorary life member, the first woman to be so honored by that organization. In 1952, at the age of seventy, she made an encouraging presentation at Dorothy Buell's house to the new Save the Dunes Council. Shortly thereafter, as the council was struggling to get enough donations to purchase Cowles Bog, Bess Sheehan emptied the savings account of the defunct National Dunes Park Association and donated its $751.68 to the Save the Dunes Council, thus assuring the purchase of that significant wetland.[14]

The Sheehans had one child, Frank Jr. Bess Sheehan died on April 24, 1968, in Winter Park, Florida. She was buried at Riverside Cemetery in Dowagiac, Michigan.

DOROTHY BUELL:
FOUNDER AND PRESIDENT OF THE
SAVE THE DUNES COUNCIL

Dorothy Richardson Buell was born in December of 1886 near Lake Michigan at Neenah, Wisconsin. She was schooled in Wisconsin Progressivism through her work with the community arts movement and graduated from Lawrence College in 1911 with a bachelor's degree in arts and oratory.[15] In 1917 she performed in the *Dunes under Four Flags* pag-

Dorothy Buell (1886–1977). *Save the Dunes/Calumet Regional Archives.*

Dorothy Buell, far left, looks at all the mayors and other officials gathered to support a national park in the Dunes. *From left*: Representative Ray Madden, George Chacharis of Gary, Richard J. Daley of Chicago, Walter Jeorse of East Chicago, Senator Paul Douglas, Mary Bercik of Whiting, Senator Alan Bible, National Park Service Director Conrad Wirth, and Interior Secretary Stewart Udall. *Calumet Regional Archives.*

eant and married James H. "Hal" Buell, a mechanical engineer. Following his career, they lived in various places, eventually moving to Ogden Dunes.

In 1949, the Buells visited White Sands National Monument in New Mexico and noted that its beauty was inferior to the dunes in Northwest Indiana. On the way home, they stopped for dinner at the Hotel Gary where they saw a sign in the lobby announcing an upcoming meeting of a group to save the sand dunes. Dorothy said, "This is what I've been thinking of. Let's go."[16]

That meeting was organized by the Indiana Dunes Preservation Council, but little progress was made right away. It was in 1952 that Buell attended a meeting of the Chicago Conservation Council at which she suggested that perhaps the women of Indiana could take the lead—as Bess Sheehan had thirty years earlier. When asked if she would assume leadership of that effort, she gave it serious thought and agreed to do so. As someone who had earlier worked in community theater, she had experience in organizing groups.

So it came to be that on June 20, 1952, Dorothy Buell gathered together a group of twenty-one women at her home and discussed the need to preserve the Dunes. Bess Sheehan attended that first meeting and told about her struggle to get the state to create the state park. Buell noted that the purpose of this new group was not to fight the proposed port, but to add five miles of unspoiled lakeshore to the state park. She later said in an interview with the press, "We are prepared to spend the rest of our lives, if necessary, to save the Dunes."[17]

The organization eventually grew to more than two thousand members, by then both men and women. Its scope expanded as well, when it was decided that the Dunes should be a national park and that the West Beach area west of Ogden Dunes should be included.

Dorothy Buell was sixty-five years old when the Save the Dunes Council was created, and she led it through some rough years. The organization became political, but not partisan; members included both Republicans and Democrats, and opposition included both Republican and Democratic officeholders. She encouraged Illinois senator Paul Douglas to join the movement and sponsor bills in Congress to establish a national park in the Dunes.

Buell's training in oratory served her well. For sixteen years she served as the council's president and was soon known as a witty and eloquent advocate for the Dunes. When the president of National Steel said that he preferred jobs to picnics, she responded, "We ask, why is it not possible to have jobs *and* picnics? Surely this is the viewpoint of a humanitarian."[18]

The struggle is well documented in J. Ronald Engel's book *Sacred Sands*. Suffice it here to say that as Congress approved funding for the Port of Indiana at Burns Harbor in 1966 it also authorized the creation of the Indiana Dunes National Lakeshore. Save the Dunes Council members celebrated, but did not quit. It took many more years to get funding authorized and the park boundaries to where they are today. Dorothy Buell retired from her presidency in 1968 and was succeeded by Sylvia Troy. The organization is still an advocate for the Dunes, fighting threats to this habitat as they arise.

In 1966, the year the national park was created, she was given two awards: Lane Bryant cited her for volunteer service and the National Wildlife Federation and the Sears-Roebuck Foundation named her Indiana conservationist of the year.[19] In 1968 Dorothy and Hal moved to Palo Alto, California, where their son Robert lived. Dorothy returned to the Dunes in 1972 for the formal dedication of the Lakeshore. Hal died in 1970 and Dorothy passed away in 1977. They are buried in Neenah. Memorial contributions in her name were requested for the Save the Dunes Council. Congress honored her in 1992 by naming the Lakeshore's visitor center the Dorothy Buell Memorial Visitor Center. In 2006, the name was applied to the new center that serves both Porter County tourism and the National Lakeshore.

PAUL DOUGLAS:

THE SENATE'S ADVOCATE FOR THE DUNES

Paul Douglas was born in Massachusetts in 1892 but was reared in a wilderness camp near Moosehead Lake, Maine. In his words, "I gained a certain serenity of spirit from the woods and mountains, along with a basic faith in the goodness of the earth."[20] He earned a PhD from Columbia University, and in 1920, at the age of twenty-seven, took a job teaching at the University of Chicago. In 1931 he married Emily Taft, daughter of sculptor Lorado Taft. She was a political activist who was later to serve as an Illinois member of the US House of Representatives.

Before being elected senator, Paul Douglas was a member of the Chicago City Council. During World War II, at the age of fifty, he served in the Marines, ending his military career as a lieutenant colonel.

Soon after arriving in Chicago, he starting spending time camping in the Indiana Dunes. On occasion he spent his nights there with Prairie Club members. After Paul and Emily were married, he built a cottage in Dune Acres from

Paul Douglas (1892–1976).
US Senate Archives.

which he could see the lake. It was at that cottage that he liked to spend his free time in the summer. Sadly, he had to sell the cottage to make money for his 1948 campaign for the US Senate.[21] That campaign was successful and Senator Douglas eventually served three terms, becoming well known for his promotion of progressive causes. He appeared on the cover of the January 22, 1951, issue of *Time* magazine, in which an article about him was titled "The Making of a Maverick."

It was a Prairie Club member who recommended Senator Douglas to Dorothy Buell as someone who loved the Dunes and might have the power to save them. Uncomfortable with the idea of a senator from Illinois meddling in an Indiana affair, Douglas tried to convince Indiana senator Homer Capehart to lead the effort to create a national park in the Dunes, but Capehart wasn't interested. Douglas then took on the challenge. He was later quoted as saying, "Until I was thirty, I wanted to save the world. Between the ages of thirty and sixty, I wanted to save the country. Since I was sixty, I've wanted to save the Dunes."[22]

Senator Douglas became a strong advocate for the park, introducing his first bill to create it in 1958. He also arranged for Senator Alan Bible and Stewart Udall, the secretary of the interior, to travel to Indiana to see the Dunes for themselves (see also chapter 7). As Douglas continued

Senator Paul Douglas alone on the beach. *National Park Service.*

to introduce bills and tried to convince others to join his cause, his opponents called him "Indiana's third senator." After years of overcoming agonizing hurdles, Congress in October 1966 finally approved legislation authorizing both a port and the park. Senator Douglas's bid for reelection the following month failed.

Paul Douglas then returned to teaching. A stroke in the early 1970s caused him to retire, and he died in 1976. But his efforts were not forgotten. In 1980 Congress authorized the Paul H. Douglas Center for Environmental Education, to be built at Miller Woods. It was dedicated in 1986.

LYNTON KEITH CALDWELL:
FATHER OF THE ENVIRONMENTAL IMPACT STATEMENT

Keith Caldwell (as he was known to his friends) grew up in Hammond, Indiana, and often visited the Indiana Dunes. He was born in 1913 in Iowa, but when his father was selected to be the superintendent of the

Keith Caldwell, far left, with Thai graduates of Indiana University, June 8, 1959.
University president Herman B Wells is third from the right. *Indiana University Archives.*

Hammond Public Schools, the family moved to that city. (Hammond's
Caldwell Elementary School is named after Keith's father.) Eventually
they purchased a house on Forest Avenue, half a block from the Illinois
state line.

Keith learned to appreciate nature and the environment while living in Northwest Indiana and visiting the Dunes—particularly Indiana
Dunes State Park. He later recalled his father's fondness for the Dunes
area, which in part led to his own passion to keep it safe. This appreciation became the foundation for his life's work and his contributions to
sustainable development worldwide. Another favorite place to go was
Chicago's Field Museum, where he was fascinated by exhibits of dino-
saurs and ancient cultures.[23] He graduated from Hammond High School
in 1930, not long after the start of the Great Depression.

In 1934, at age twenty, Keith completed his bachelor's degree (with
honors) in English at the University of Chicago and got a job teaching
at Whiting Junior High School. Finding junior high teaching neither

Caldwell at his home, Cedar Crest, in April
1972. *Indiana University Archives.*

easy nor fulfilling, he tried teaching at the high school. Then in 1936 he
started a summer graduate program at Harvard, returning to Northwest
Indiana in the fall to teach. Once home, he enjoyed weekend trips out
to the Indiana Dunes.

Appreciating the stimulating atmosphere of Harvard, Keith resigned
his teaching position and returned to Cambridge to complete a master's
degree in history. That fall he returned to Hammond and enrolled in a

doctoral program in public administration at the University of Chicago. In August 1939, the following year, he got his first full-time job teaching at Indiana University's Calumet Center in East Chicago. It was at a dedication of the center's new building on Indianapolis Boulevard that he met IU president Herman B Wells. The two men were to become good friends. The next fall he was named the first full-time executive secretary of the university's South Bend extension. That December Keith married Helen Walcher, a secretary at the NIPSCO offices in Hammond.[24] They raised two children.

In 1943 Keith Caldwell received his PhD in political science. By this time he had joined both the Izaak Walton League and the Audubon Society. In 1944 he started a chapter of the latter group in South Bend and served as its first president. Just four months later, however, he left South Bend to take a position in Chicago.

Thirteen years later Caldwell, now Professor Caldwell, returned to Indiana University, this time to the Bloomington campus where he accepted a faculty position in the Department of Government. A civics expert with a strong attachment to environmental causes, he joined The Nature Conservancy and became the first chairman of its new chapter. He actively supported the efforts of the Save the Dunes Council and on occasion gave presentations promoting the proposed national park at the Dunes.

Caldwell's 1963 *Public Administration Review* article "Environment: A New Focus for Public Policy?" essentially started the field of environmental policy studies.[25] The article proposed, among other things, that environmental policies needed to include rules requiring actions that would promote appropriate environmental decision-making. The timing was perfect: Rachel Carson's best seller *Silent Spring* had just been published and Secretary of the Interior Stewart Udall had already alerted President Kennedy to the nation's serious environmental problems.

Caldwell then became the leader in the area of environmental regulation. He was a cowriter of the National Environmental Policy Act (NEPA), the first law of its kind on the planet. Perhaps the most important contribution he made was NEPA's requirements for public participation, written responses to concerns, and the researching and writing of environmental impact statements before approval would be given for development projects.

He served as an advisor to the US Senate, the US Departments of Commerce, Energy, Defense, and Interior, the National Institutes of Health, and the United Nations. In 1972 he and IU Northwest geology professor Mark Reshkin were instrumental in creating the School of Public and Environmental Affairs (SPEA) at the Indiana University Bloomington and Northwest campuses.

Walter Helminski (1923–1998).
Walter Helminski Jr.

Keith Caldwell officially retired in 1984, but like many professors, he didn't stop working after his retirement. Indeed, he continued writing and lecturing until about 1990. His last book, *The National Environmental Act: An Agenda for the Future*, was published in 1988. In 2001, seventeen years after his "retirement," he was the recipient of Indiana University's Distinguished Service Award. Caldwell died at the age of ninety-two on August 15, 2006, at his home in Bloomington. His primary legacy is the National Environmental Policy Act, with its vital component, the environmental impact statement.

WALTER HELMINSKI:
THE REGION'S RECYCLING PROPHET

For many years, Walter Helminski was a persistent but lonely voice urging Calumet Area communities to find better ways to dispose of its solid waste. As described in an editorial written in 1990 by the *Times* newspaper, "He was a nag. Our conscience. He was a man on a mission. Ahead of his time. Today, thanks to Helminski and other far-sighted individuals who understand the importance of a sound environment, many communities in the area have begun recycling programs. But others have not—and much work remains to be done."[26]

Helminski was an outside plant engineer for the Illinois Bell Telephone Company and had a longtime interest in the natural world. At the age of fifty-three, he founded a nonprofit organization called Region Recycling and convinced others to join him in his efforts to establish either a town-wide or a county-wide recycling program. Region Recycling set up one of the area's first many-item recycling centers on the former Nike site on south Columbia Avenue in Munster. Helminski then started writing a weekly column about recycling and the problems of solid waste disposal. Finally, in 1990, largely because of his efforts, Munster became the first Calumet Area community to establish a town-wide curbside collection service, a program since copied by many other towns and cities in the area.

The *Times* concluded its editorial by saying, "Heroes come in all shapes and sizes. Helminski is a hero. The area owes him a tremendous debt of gratitude." Walter's wife Betty passed away in 1994. Walter died on March 16, 1998, at the age of seventy-four.

Environmental Education Opportunities

15

In order to have a relationship with nature and a desire to take care of it, students have to take the time to get to know it, experience it, and then they can love it.

Mighty Acorns, *Curriculum Guide*

FORMAL EDUCATION ABOUT THE ENVIRONMENT OF NORTH-west Indiana began when Henry Chandler Cowles started to bring his University of Chicago students to see the unique biological assemblage there. Before the South Shore Railroad was established, the only rail transportation to the Dunes was the Lake Shore and Michigan Southern Railway, and so his field trips were near its Miller, Dune Park, and Chesterton stations.

Cowles wrote that Chesterton was "the most interesting place in the vicinity of Chicago, since it shows types of nearly all plant societies, all phases of the river series . . . , all stages from pond to prairie, all types of dune activity."[1] Dune Park, he wrote, "is [by] far the best place to study living dunes in all phases."[2]

UNIVERSITY EDUCATION

Today all of the major local universities and Ivy Tech Community College have courses in environmental education. Faculty members are often advisors, consultants, volunteer workers, or board members of environmental nonprofit organizations. Several courses require that students gain some practical experience.

IU Northwest offers a bachelor of science degree in environmental science and also has its School of Public and Environmental Affairs. It offers advanced courses in fungi, plant diversity, regional ecology, con-

Henry Chandler Cowles and his students relaxing during a
study trip to the Dunes. *National Park Service.*

servation biology, biological chemistry, environmental policy analysis,
and environmental health. Purdue Calumet offers an environmental science minor and an ecology concentration with its biology degree. It also
offers advanced courses in aquatic ecology, invasive species ecology, and
climate change. Valparaiso University offers a degree in environmental
science. Calumet College has a course in environmental science. Ivy
Tech offers courses in environmental management and water quality
management, and an introductory course in environmental law.

In addition to courses on the environment, four of these institutions
have programs that focus on local environmental issues. For instance,
Purdue Calumet has both an office of the Illinois-Indiana Sea Grant
and its Purdue Calumet Water Institute. IU Northwest has its Center
for Regional Excellence. They all work closely with local organizations.

The Northwest Indiana Restoration Monitoring Inventory (NIRMI)
is a project that measures how components of a restoration have pro-

gressed over time. This project uses standardized field methods to monitor restorations and provides data and analyses to maintain successful restorations. With data collected since 2010 students and faculty have developed the NIRMI Restoration Index, the first such index in the country, in which positive numbers indicate increasing ecological value and negative numbers indicate a decrease. Undergraduate interns are hired from the local colleges and universities; they assist with all aspects of the project and are especially important to NIRMI fieldwork. The students learn methods of plant identification and field ecology techniques as well as techniques in bioinformatics, GPS/GIS tools, and web-based integration of data.[3]

The Great Lakes Innovative Stewardship Through Education Network (GLISTEN) uses a service-learning approach to education in which students use community-service and research activities to supplement classroom instruction.[4] Undergraduate students from IU Northwest, Valparaiso University, and Ivy Tech volunteer to help environmental agencies and organizations such as Save the Dunes and Shirley Heinze Land Trust restore nature preserves, gather information about stream water quality and combined sewer overflows, assist in watershed restoration, and monitor restoration projects (through NIRMI), and also help create the Northwest Indiana Water Trails Inventory.

The Illinois-Indiana Sea Grant, housed at Purdue Calumet, serves the Chicago metropolitan area while fostering, among other things, restoration efforts that affect Lake Michigan water quality. It funds research, publishes papers and educational materials, and performs outreach services. It brings together many stakeholders in order to improve the environment and economy of the area.[5] Many field experts also work with its River Restoration workshop series, and its BeachWatch project provides information regarding Northwest Indiana beach closures.

Valparaiso University (VU) also has a long history of working with local school districts. Recently it sponsored its Initiative for Schools, Industries and the Sciences, in which science teachers from area schools were invited to summer workshops to learn how to use sophisticated instruments in standards-based lessons; the university then lent that equipment to the teachers for the school year. In a later project, the university obtained a grant from the EPA for its Building Bridges for Environmental Stewardship program, in which VU students work with local middle school students in school classrooms, out in the field, and at the university science labs.

ENVIRONMENTAL EDUCATION IN THE SCHOOLS

With both children and adults spending more time indoors than their grandparents did, outdoor experiences and education are more important than ever. Fortunately, there are many ways to learn about the environment of the Calumet Area and its sustainability.

Indiana academic science standards for elementary and middle schools include many references to the environment. This begins in the first grade, when students study key principles of nature and the relationships of living things to their environment. Environmental education is then expanded upon in the higher grades, especially in grade four. In middle school, understanding how environments react to change is an important standard in grades seven and eight. Then in high school, the process standard "Explain how scientific knowledge can be used to guide decisions on environmental and social issues" is a part of every science course. Environmental issues play an especially big role in high school biology, a course that most Indiana students take.

Michigan City's Krueger Middle School has a habitat restoration project area adjacent to the school. Its teachers and students have worked with NIRMI interns to establish the monitoring of its restoration efforts. NIRMI has also worked with the Charter School of the Dunes, assisting faculty with the development of its restoration project on seven acres of land.

DUNES LEARNING CENTER

Dunes Learning Center, a residential provider of educational experiences in nature, is located within the Indiana Dunes National Lakeshore, not far from where Henry Chandler Cowles studied the plant life of the Dunes and developed his theory of succession. It has ten cabins, each with four bunk beds, a bathroom, and showers. Schools from four states take advantage of the center's programs, such as Frog in the Bog for ele-

mentary school students or Winter Ecology in the Dunes, Dune-SCOPES, and Science Olympiad Team Training for middle and high school students. These programs run through the school year and are designed to include at least one overnight stay so that students get an in-depth experience. Dunes Learning Center also runs a full summer program of weeklong camps for kids ages six to seventeen, as well as Critter Camp, a day camp with a nature theme, for those ages six to eight.[6]

The goal of DLC programs is not to have children simply memorize names of trees and bugs, but rather to help them develop an understand-

The Dunes Learning Center campus at the Indiana Dunes National Lakeshore.

ing of the relationships between living organisms within a natural community. Since its establishment in 1998, Dunes Learning Center has served more than one hundred thousand children, many of whom spent their first night away from home at the center. Nearly all of them saw their

first hawk, heard their first frog, identified their first tree, or had their first in-depth discussions about the environment there. Others first learned about food waste, energy savings, deer populations, sustainability, and shoreline erosion. Even for some local students, their days at the center gave them their first view of Lake Michigan.[7]

Adult education programs at the center, for teachers and others who may be interested, are run by staff from the Indiana Dunes National Lakeshore, the Great Lakes Research and Education Center, and Dunes Learning Center. These workshops on various topics include an array of hands-on activities within the National Lakeshore.

SHIRLEY HEINZE LAND TRUST

Shirley Heinze Land Trust was founded in 1981 with the goal of preserving environmentally significant places in Northwest Indiana. In its thirty-five years, it has purchased or been given close to 2,000 acres of natural areas in seventeen locations across Lake, Porter, and LaPorte Counties. The trust's mission is twofold: to protect Northwest Indiana habitats and ecosystems by "acquiring, restoring, and protecting environmentally significant landscapes," and to educate people about how land conservation protects nature and enriches our lives.[8]

Many of the Heinze Trust's preserves have public trails and parking areas that are open from dawn to dusk every day. Permitted daily activi-

Facing, top: "Look what I've got!" Students looking at a salamander. *Dunes Learning Center.*

Facing, bottom: Frog in the Bog students on a hike to the beach on the second day of the program. *Dunes Learning Center.*

Jim Erdelac, education and volunteer manager at Shirley Heinze Land
Trust (*right*), on a Michigan City Boys and Girls Club hike, shows children
the annual growth rings on a tree stump. *Shirley Heinze Land Trust.*

ties include hiking and cross-country skiing (with at least two inches of
snow) on the designated trails, birding, photography, and nature study.
Dog walking is allowed as long as the dogs are on a lead and any feces
are picked up and carried out.

The Heinze Trust's properties include areas that contain examples of
most of Northwest Indiana's natural ecosystems. These include prairies,
savannas, woodlands, marshes and swamps, ponds and fens, riverbanks,
dunes, and the quite rare dune-and-swale habitat. Five of the trust's properties are dedicated Indiana state nature preserves.[9]

MIGHTY ACORNS

Fostering a Personal Connection to Nature through Stewardship

The Nature Conservancy and the Forest Preserve District of Cook
County created the Mighty Acorns program in 1993. It was developed
and tested by classroom teachers and other volunteers. Chicago Wilder-

ness, a large coalition of conservation agencies and organizations,
adopted the program in 1998 as a
model environmental education
program and then secured a grant
so that it could be expanded into
neighboring counties. Chicago
Wilderness also identified the
program as one of the principal

components of its Leave No Child Inside initiative, "a comprehensive
effort that aims to develop the region's next generation of environmental
leaders."[10] Dunes Learning Center initiated the Mighty Acorns program

Students from the Charter School of the Dunes collect seeds from
native plants at the Ivanhoe South Nature Preserve in west Gary,
managed by Shirley Heinze Land Trust. *Dunes Learning Center.*

in Indiana. Shirley Heinze Land Trust added Mighty Acorns to its education programs in 2009.

Mighty Acorns is school based. It is aligned with both Illinois and Indiana academic standards and enriches classroom curricula with actual hands-on restoration work at a nature preserve or other natural area, usually near the school. If no such area is located near the school, the students are bused to such an area. In a win-win arrangement, students learn about nature as they care for "their" plot of land, and the outdoor area is made healthier as a result of their restoration work.

Dunes Learning Center launched its Mighty Acorns outreach program with three schools in Gary and Hobart in the fall of 2000. A lead naturalist from the center goes out to the schools and works with the teachers and students there and at their chosen restoration area.

During the school year, Mighty Acorns students make at least three work-study trips to their natural area. Each visit to their preserve has a different focus. In the fall they identify native plants and collect some of their seeds; in winter they may remove nonnative shrubs; then in the spring they sow the native seeds they collected the previous fall. They may also start some of the seeds in their classrooms. Topics in the classroom include the interdependence of plant and animal life, competition for resources, diversity, and adaptation.[11] The program gives students a better sense of place, an understanding of the needs of living things, and a feeling of pride for improving their neighborhoods.

Chicago's Field Museum is the partnership coordinator for all Mighty Acorns programs throughout the Calumet Area. The staff there trains all the partner leaders and assists with teacher trainings as well.[12]

Mighty Acorns Summer Camp

Students who participate in Mighty Acorns in their schools are then invited to spend a week in the summer at the Mighty Acorns Nature Camp at Dunes Learning Center. During their five days and four nights at the camp, the students explore the many different ecosystems in the Dunes, learn about wild animals and their habitats, take part in restoration and stewardship projects, and, of course, sing campfire songs (including the Mighty Acorns song) and make s'mores.

Shirley Heinze property assistant Peg Mohar leads a group of Mighty Acorns students from South Haven Elementary School out to discover strategies plants use to spread their seeds. *Shirley Heinze Land Trust.*

Mighty Acorns summer campers at Dunes Learning Center study macroinvertebrates living in the Little Calumet River. *Dunes Learning Center.*

STATE AND NATIONAL PARK PROGRAMS

Both Indiana Dunes State Park and the Indiana Dunes National Lakeshore have excellent and extensive interpretation programs for children and adults, and both have wonderful facilities for their education programs. Both parks sponsor seasonal programs, such as Maple Sugar Days at the Chellberg Farm and workshops funded and conducted by Friends of the Indiana Dunes. Both sponsor programs for adults that extend beyond the boundaries of the parks themselves. Such programs include field trips in the fall to see thousands of sandhill cranes arrive at the Jasper-Pulaski Fish and Wildlife Area (which the Indiana Department of Natural Resources called "one of Indiana's greatest wildlife spectacles"[13]) and illustrated lectures followed by geology-based tours of Northwest Indiana.

Indiana Dunes State Park Programs

Indiana Dunes State Park may have been the first agency in Northwest Indiana to offer conservation and environmental programs. The park schedules special programs throughout the year for both day visitors (both adults and schoolchildren) and campers. Many of its programs and trails begin at its recently restored, kid-friendly Nature Center, which has many exhibits of its own.

Some Indiana Dunes State Park programs cover living things such as beavers, birds (including owls, waterfowl, and the wild turkeys that inhabit the park), butterflies, bats, bunnies, and ant lions. Other programs concern physical attributes of the park, such as its creek, sand dunes, blowouts, and wetlands. Each August, there is a nighttime sky-gazing program during the annual Perseid meteor shower. Some programs are held in the Nature Center, while others are conducted during hikes. Winter hikes can include cross-country skiing, though that, of course, is weather dependent.[14]

The 3 Dune Challenge is a rugged 1.5-mile hike that climbs the three tallest dunes in the park. Those who complete the hike can pick up a free sticker and postcards or purchase a commemorative T-shirt or hoodie.

The Nature Center at Indiana Dunes State Park.

National Lakeshore staff and volunteers explain how American Indians made maple sugar. *Photo by Jeff Manuszak, National Park Service.*

Indiana Dunes National Lakeshore Programs

Unlike many of the other national parks, the Indiana Dunes National Lakeshore's original mission statement included an education component, and thus education has been an important part of the Lakeshore's program since the mid-1960s. The park has an interpretive staff that provides instruction for school-based field trips to the Dunes and summer and evening programs for families. Many of the educational programs are held at the Paul H. Douglas Center for Environmental Education, which opened in 1986.

The Lakeshore has educational programs for children and adults. Every year thousands of students participate in formal education programs at various locations. Many programs, of course, include information about the various ecosystems within the park, but cultural and historical programs are held as well. Teachers may request program packets in advance so that they and their students are prepared for their trips. There are ranger-led programs throughout the year, but many people like to come and explore on their own. Kiosks located in various places feature educational information about the park's flora and fauna, and the *Singing Sands* park newspaper includes a park map and a calendar of events.

SOLID WASTE DISTRICT PROGRAMS

All three of the Calumet Area's solid waste management districts have extensive and free K–12 educational programs that are aligned with Indiana state academic standards. In addition, each of the three districts has developed some very interesting and innovative educational programs.

Lake County has two Environmental Educational Centers, one in Merrillville and one in Hammond, which are used as field-trip sites for school groups. The Hammond site is also the location of the district's summer camp programs. In addition, the district has an "Enviromobile," which is taken to schools (and, in the summer, to libraries, scout groups, and church groups) in order to conduct some of its ten hands-on inquiry-

Facing, top: An orientation session for a class at the Paul H. Douglas Center for Environmental Education. *National Park Service.*

Facing, bottom: A ranger-guided butterfly hike. *National Park Service.*

The Environmental Education Center in Hammond.

Porter County District educator Abe Paluch helps kindergarten students understand the anatomy of a red wiggler worm as part of the Wonderful Worms program, one of many free environmental programs offered to schools. *Recycling & Waste Reduction District of Porter County.*

Teaching about the contribution worms make to compost.
Lake County Solid Waste Management District.

based activities. Also going out to schools is a performance group called the Green Team, which puts on a high-energy show for school assemblies that emphasizes the three environmental Rs (reduce, reuse, recycle).[15]

Porter County staff makes presentations to community groups and conducts interactive education programs at schools throughout the county, annually reaching more than six thousand students. It has also offered adult workshops on rain barrels, healthy homes, and backyard composting. In 2014, the district launched the Porter County Master Recycler Program, a comprehensive multiweek class for adults, the first of its kind in the state of Indiana. The program features guest speakers and field trips for participants who want to learn how to reduce their impact on the planet.[16]

LaPorte County's website features teacher lesson plans, how- to articles, green games, craft ideas, and articles about being green. To demonstrate that recyclables are not dumped with garbage into landfills, it also includes a video tour of the Diversified Recycling plant in Homewood, Illinois, which shows how recyclables are sorted using huge conveyor belts. The district also provides a very helpful downloadable Visual Recycling Guide.[17]

OUTDOOR EDUCATION THROUGH SCOUTING

All scouting organizations emphasize outdoor education. Boy Scouts of America offers specific education and experiences through more than 130 merit badges in 14 subject areas, which include fish and wildlife management, insect study, nature, plant science, reptile and amphibian study, and weather.[18]

Girl Scouts of the United States of America is divided into six grade levels: Daisies, Brownies, Juniors, Cadettes, Seniors, and Ambassadors. Each level has patches that can be earned. Among the environmental patches are ones for bugs (for Brownies); geocaching, animal habitats, and flowers (for Juniors); trailblazing, animal helpers, and trees (for Cadettes); the sky and the relationship between humans and animals (for Seniors); and water (for Ambassadors).[19]

Preservation and Restoration of Natural Areas

16

This chapter describes several, but certainly not all, preserved lands in the Calumet Area. For more than a century, the prevailing attitude in the Midwest about open lands was that they were just (yet) "undeveloped." Dry, flat, open areas were ready to become profitable. Empty lots in a residential neighborhood were "vacant," an almost negative term that practically invited development. Wetlands and hills inhibited both farming and urban development. The wetlands had to be drained first—and many of them were—and hills and sand ridges had to be removed. What made this process not only easy but also profitable was that in northern Indiana all the hills and ridges were composed of sand or clay as opposed to solid rock (as they are in much of southern Indiana). Removing sand dunes or ridges with steam-based equipment was usually not financially impossible. The sand or clay could then be sold. Filling in or draining wetlands was quite possible as well.

In 1900, candidate Teddy Roosevelt emphasized conservation in his campaign for vice president. Also that year, the Chicago Municipal Science Club first proposed the setting aside of scenic areas of Cook County—setting in motion the establishment of the Cook County Forest Preserves. Daniel Burnham's famous *Plan of Chicago*, completed in 1908, had among its primary goals the reclamation of the lakefront for the public. He wrote, "The Lakefront by right belongs to the people. Not a foot of its shores should be appropriated by individuals to the exclusion of the people."[1] Chicago's Prairie Club became a major force in conservation in 1911.

Indiana Dunes State Park by South Shore Line, 1929.
Lithograph by Raymond Huelster, held at the Calumet Regional Archives.

A crowded beach at Indiana Dunes State Park, June 9, 1968. *Save the Dunes.*

INDIANA DUNES STATE PARK

Indiana Dunes State Park was one of the first tangible results of the "save the Dunes" efforts of the early twentieth century, thanks to the extraordinary efforts of Bess Sheehan described in chapter 14. After the Indiana General Assembly authorized a state park on the shores of Lake Michigan, Samuel Insull, owner of the South Shore Railroad, personally donated funds so that electric lines could be installed underground. The railroad then often featured depictions of the new state park on its advertising posters. Chicagoans were, of course, encouraged to take the train to see the Dunes.[2]

The park, with its three miles of beach and about three and a half square miles (2,182 acres) of land, opened on July 1, 1926, four years ahead of the original schedule.[3] In its first three months, nearly sixty-three thousand people came to visit. The limestone pavilion was built and opened in 1929, and two years later the Dunes Arcade Hotel, with forty-four rather small guest rooms, opened just west of the pavilion. The hotel was demolished in 1972.

This state park soon became one of the most visited in Indiana. By the mid-1950s, attendance had skyrocketed, and on hot summer weekends

The new bird-watching tower.

the park would reach its visitor capacity and officials would turn away cars at the entrance. This is no longer done, but now when the park is at capacity, cars may be held up at the entrance and allowed in only as others leave.

Besides the beach, the state park has a beautiful landscape containing woodlands, wetlands, and blowouts. In addition, it has an inspiring history. The eastern two-thirds of the state park consists of the nature preserve, which includes the three tallest dunes—Mounts Tom, Holden, and Jackson—but also marsh and forest lands. The twenty-first century has seen improvements: in 2004–2005 two large and modern comfort stations were built at the campgrounds; in 2012, Dunes Creek, which had been diverted into a culvert so that the parking lot could be expanded, was reopened to the air; and a new bird-watching tower was built atop a tall foredune in 2015. And in 2010, after years of being a favorite vacation spot for Hoosiers, the park's attributes finally became recognized across the country as *Midwest Living* magazine named the park one of the Midwest's thirty-five best state parks—a recognition that has been repeated since.[4]

West Beach, one of the first properties purchased by
the National Park Service. *Kim Swift.*

INDIANA DUNES NATIONAL LAKESHORE

Indiana Dunes National Lakeshore, established in 1966 and expanded
in 1976, 1980, 1986, and 1992, is composed of approximately fifteen thousand acres. It is a unique national park in that it is located in an urban,
industrialized region, and is composed of several distinct parcels separated by residential and industrial areas. After more than 120 years of
study, beginning with University of Chicago botanist Henry Chandler
Cowles's work in 1896, more than 1,574 plant species have been found
in the park.[5] This wealth of plant species is exceeded in only two other
national parks: the Great Smoky Mountains and the Grand Canyon,
both of which are much larger than the Lakeshore. The Indiana Dunes
has the greatest degree of plant diversity in the whole upper Midwest or
Great Lakes region. Within the national park boundaries are the Lake
Michigan shoreline and sand dunes, both modern and ancient, as well as
expansive beaches, savannas, forests, marshes, bogs, and fens.

Eight million people live within a day's drive of these parks. Mark
Reshkin has called the Indiana Dunes a prototype for urban national
parks, noting that natural land management within the Indiana Dunes
area may very well demonstrate whether or not scientific and aestheti-

cally valuable areas can be preserved and sustained in heavily industrialized urban areas of the world.[6]

There are scores of small and large sites within the Lakeshore's boundaries that have undergone restoration work. The largest may be the work on the Great Marsh.

Great Marsh and Cowles Bog Restoration

According to the Indiana Department of Environmental Management, before settlement, wetlands covered up to a quarter of what is now the state of Indiana. However, by the late 1980s, as a result of ditching and other draining methods, more than 4.7 million acres, or 85 percent of Indiana's wetlands, had been lost.[7]

The lowlands from LaPorte County westward toward Illinois that were between the higher Tolleston and Calumet Shorelines once were often saturated with water and supported great biodiversity. The Little Calumet River, west of Salt Creek in Portage, was part of this area, and so were Cowles Bog and the Great Marsh, which, south of the residential areas of Dune Acres and Beverly Shores, originally extended from Burns Harbor nearly to Michigan City. The area was home to thousands of ducks, geese, and wading birds. This great diversity of plants and animals reflected a healthy ecosystem.

Wetlands used to be considered worthless wastelands. Folks believed that they had to be drained in order to make them productive. In 1850 Congress passed the Swamp Land Act, which gave federal wetlands to the states, which could then sell them to buyers who would arrange and pay for their draining. Once drained, the land would be suitable for farming. Eventually scientists and government officials realized that wetlands have real value, so in 1972 Congress finally passed the Wetland Protection Act. During the intervening 120 years, levees, ditches, roadways, and industrial, commercial, and residential buildings altered the landscape of these wet areas, which have significantly shrunk in size as a result.

By the 1980s, much of the Great Marsh had changed from an area with a great diversity of plants to one characterized by cattails and other invasive species. These nonnative plants reduced habitats required by herons, egrets, and other waterfowl as well as other animal species that need healthy wetlands to survive.[8] In 1998, the National Lakeshore, using

Backbreaking restoration work in the Great Marsh. *National Park Service.*

Cowles Bog, with a great egret and mute swans. *Shirley Heinze Land Trust.*

A great blue heron near Beverly Drive. *National Park Service.*

staff and volunteer help, began restoring the five-hundred-acre Derby Ditch section of the Great Marsh in Beverly Shores. Students from the Dunes Learning Center and various schools and scouting groups have assisted with this project.[9]

Shirley Heinze Land Trust, which has acquired eighty acres of marshland outside of the Lakeshore's boundaries, primarily north of Beverly Drive in Beverly Shores, has also begun restoration work there. The Na-

The greenhouse at the National Lakeshore where seedlings await transfer to restoration areas. *National Park Service.*

tional Audubon Society has recognized this whole area as an "Important Bird Area" and has listed it as a designated stop on the Chicago Region Birding Trail.

Cowles Bog is a special area within the Great Marsh. It was purchased by the Save the Dunes Council in the 1950s, designated one of the nation's first national natural landmarks in 1965, and donated to the new National Lakeshore in 1966.

The Cowles Bog Restoration Project is bringing new life back to the wetlands. Adult volunteers, scout and school groups, and other dedicated citizens, working with the National Lakeshore's staff, have started removing invasive plants and replacing them with native sedges, grasses, shrubs, and trees. These efforts are causing the grounds to attract a more diverse animal population, including migratory birds seeking a resting place. This work, which is also reducing and controlling runoff, is in addition helping to improve the water quality of Lake Michigan. (Incidentally, the Cowles Bog area contains the only remaining wild stand of northern white cedar in Indiana.) There is still much to be done, but already flocks of mallard and wood ducks, great egrets, and great blue and green herons, and even beavers, are making the Great Marsh their home again—some after an absence of more than one hundred years.

OTHER PRESERVED NATURAL AREAS

Since the 1960s, there has been a quiet revolution in the way Americans in general, and Calumet Area residents in particular, have come to think about the land. The Save the Dunes Council, which preceded this change of heart, had to fight hard to get newspaper coverage and congressional approval of a national park in the Dunes. But since then, and particularly since the inception of Earth Day celebrations in 1970, there has been widespread interest in open space, native species, and proper stewardship of the earth.

While it is true that parks have been important to most people for more than a century, interest in local parks and local environmental restoration activities in Duneland has increased significantly since 1970. The value of native landscapes, beyond just their open space, is being recognized by more people every year. The Friends of the Dunes' native plant sale is its biggest annual fundraiser.

The Calumet Area has seen many different forms of restoration. Some valuable efforts may just be simple cleanups, while others may involve the removal of nonnative, particularly invasive species or the planting of species native to the area. Still other restorative work may involve the removal of drainage tiles or ditches that were earlier used to drain the water off of wetlands. The Heritage Trust license plate program, administered by the Indiana Department of Natural Resources, has collected voluntary contributions from Hoosier citizens and used those funds for hundreds of restoration projects—many of which have been located in Duneland.

Nongovernmental organizations, such as The Nature Conservancy, the Izaak Walton League, Save the Dunes, Shirley Heinze Land Trust, the Wildlife Habitat Council, and others, have contributed to all of these efforts. They have purchased and restored literally thousands of acres of wetland prairie, and woodlands in all three counties of the northwest corner of the state.

The biggest industries, as well as small businesses, local park districts, schools, and private landowners, have contributed to these efforts as well. For example, there are now natural areas with walking trails on private land for the benefit of companies' employees. Taken together,

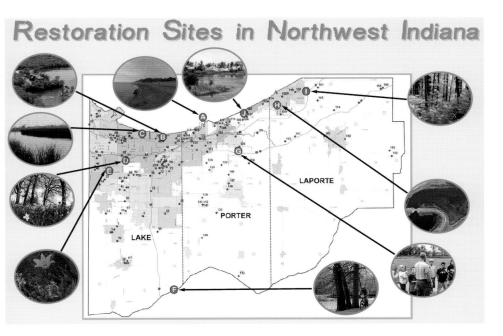

Restoration Sites in Northwest Indiana

Restoration Sites in Northwest Indiana. *IU Northwest Center for Regional Excellence.*

A. Portage Lakefront and Riverwalk
B. Marquette Park Lagoons
C. Grand Calumet River
D. Little Calumet River / Hobart Marsh
E. Hoosier Prairie

F. Grand Kankakee Marsh
G. Coffee Creek Watershed Preserve
H. Trail Creek and Striebel Pond
I. Ambler Flatwoods
J. Great Marsh

these efforts illustrate the real concerns that private citizens, as well as local businesses and industries, have about the greater environment.

A 2011 INVENTORY OF RESTORATION SITES IN NORTHWEST INDIANA

In 2006, Indiana University Northwest's Center for Regional Excellence partnered with the Northwestern Indiana Regional Planning Commission (NIRPC) and the Quality of Life Council to determine the status of Northwest Indiana restoration efforts by creating an inventory of ongoing restoration sites. They found a total of 166 sites, including those listed above.

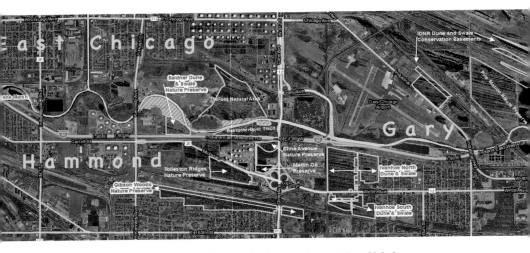

Key restoration areas in Northwest Lake County. Areas with red labels indicate sites with public access. *The Nature Conservancy.*

RIVERINE AND DUNE-AND-SWALE PRESERVES IN NORTHWESTERN LAKE COUNTY

Dune-and-swale topography is a rare landform composed of roughly parallel low sand ridges separated by often wet swales. The ridges here were made by the waves of Lake Michigan during the lake's later Tolleston stage when sand was washed up on the beach during high-water stages. Not exactly coincidentally, the landform was formed in the same part of Northwest Indiana in which its four lakeshore cities were located and largely where the heaviest industry was sited. Yet after 150 years of development, surprisingly there are pockets of the original dune-and-swale landscape that have not been disturbed.

Gibson Woods

The first of these properties to be preserved was the long and narrow 131-acre Gibson Woods property that was owned, but never developed, by the Michigan (later New York) Central Railroad. It was purchased by The Nature Conservancy in 1980 and then sold the following year to Lake County Parks. On November 8, 1981, the park was dedicated by the state of Indiana as a nature preserve. Park staff have found on-site more

The nature center and boardwalk at Gibson Woods.

than three hundred species of plants, several of which in Indiana are listed as threatened or endangered. Gibson Woods has a nature center with a bird-viewing room and a full-time staff.[10]

THE GRAND CALUMET RIVER NATURE PRESERVES OF HAMMOND AND EAST CHICAGO

As can be seen on the aerial photograph of key restoration areas in Lake County, there are numerous preserves along the Grand Calumet River. They are owned by several different organizations.

Roxana Marsh in East Chicago

Before the sediment remediation and restoration projects, Roxana Marsh was a wasteland overrun by invasive phragmites (the light brown area in the 2011 photograph of the marsh), offering little to no ecological benefit to birds, turtles, or fish. Students from local schools helped plant wetland plants at the marsh, and after just one growing season, Roxana Marsh was full of life, as butterflies and bees buzzed atop black-eyed Susans and

The phragmites-infested Roxana Marsh in 2011, before restoration.
Photo by Scott Ireland, US Environmental Protection Agency.

Roxana Marsh in 2013, after restoration.
Photo by Scott Ireland, US Environmental Protection Agency.

Sand coreopsis in bloom at the Seidner Preserve.

other native plants. It has become a beautiful and healthy community amenity as well as a resting spot for migratory and nesting birds.

The EPA, the Indiana Department of Natural Resources, IDEM, and the US Fish and Wildlife Service all participated in this project, which remediated six hundred thousand cubic yards of contaminated sediment (including the adjacent section of the Grand Calumet River). Contaminants included heavy metals, oil, grease, PAHs, and PCBs.[11]

SEIDNER DUNE AND SWALE, BEEMSTERBOER, AND DUPONT

The first forty-three acres of the Seidner Dune and Swale property south of the Grand Calumet River were purchased by Shirley Heinze Land Trust in 1997. Seven adjacent acres were later donated by Save the Dunes. The preserve contains both dune and swale and riverine marsh areas. The trust has done extensive restoration work there.

The land had been zoned for industrial purposes but was not developed, except for a gravel road and two rail spurs roughly built upon two

A great egret foraging for fish in the restored Grand Calumet River north of the Beemsterboer property, a sign of the improving health of the river. *Photo by Susan MiHalo, The Nature Conservancy.*

of the sand ridges. This state-dedicated nature preserve was named after Robert and Bette Lou Seidner, who in 1981 provided the initial funds to establish Shirley Heinze Land Trust.

In 2009, cleanup began on the section of the Grand Calumet River between Kennedy and Cline Avenues. As the cleanup of this portion of the river was completed, The Nature Conservancy coordinated the start of work on an adjacent marsh and what used to be an extensive dune-and-swale black oak savanna. This area includes the Beemsterboer property south of the river and the DuPont conservation easement north of it. Restoration of emergent marsh and wet prairie at Seidner is part of this project.

Since the restoration efforts began, Canada geese, turtles, ducks, egrets, herons, red-tailed hawks, deer, foxes, and beavers have been spotted there.[12] Bald eagles have also been sighted on riverside treetops.

DUNE AND SWALE IN WESTERN GARY

The Indiana Department of Natural Resources, The Nature Conservancy, and Shirley Heinze Land Trust have all acquired and restored

Lupine blooming near one of the wet swales. *The Nature Conservancy.*

Ivanhoe South Preserve. *Shirley Heinze Land Trust.*

dune-and-swale landscapes in the less developed western part of the city of Gary. Ivanhoe Dune and Swale Nature Preserve is a ninety-two-acre parcel north of Fifth Avenue (US Route 20) and west of the houses on Hobart Street. It is owned and maintained by The Nature Conservancy. There is no parking lot and there are no trails.

The dune ridges have black oak trees; however, they are far enough apart that sunlight can still shine on the prairie grasses and wildflowers that form the understory, such as puccoon, spiderwort, and prairie phlox. Some of the swales are buttonbush swamps, while others are sedge meadows. Yellow water buttercups, which grow in profusion in the swales, are a favorite plant of many species of butterflies that frequent the area.

The fifty-acre Ivanhoe South Preserve south of US 20 is actually a collection of dozens of individual city lots that were sold years ago to would-be home builders but never developed. The lots have been purchased one or two at a time by Shirley Heinze Land Trust as they have become available at tax sales; many other lots were donated by landowners. In 2012, the city of Gary graciously ceded to the trust the strips of land once designated as streets. This preserve has a hiking trail, which begins at its parking lot on Colfax Street.

OTHER PRESERVES IN HAMMOND

Hammond Lakefront Park and Bird Sanctuary

Part of the Lake Michigan bird flyway, the Hammond Lakefront Park and Bird Sanctuary is located on the lakefront, just a short distance east of the Indiana-Illinois state line. Described in *Chicago Wilderness* magazine as one of Indiana's best bird-watching sites, it occupies just sixteen acres of land, nine of which compose the actual sanctuary. Yet, during the spring and fall migration periods, largely because it's the only available acreage in that general vicinity, one may see hundreds of species of songbirds, waterbirds, and even birds of prey such as snowy owls and peregrine falcons.[13]

The park and sanctuary, which used to be a concrete dumping ground, are owned and managed by the city of Hammond. They could be called a former brownfield, but trees grew among the rubble, and so this island of green, nestled between the Hammond Marina and a former generating

The Hammond Bird Sanctuary on the edge of Lake Michigan.

station, was preserved by the city and not incorporated into the marina complex. The park has a footpath and benches for those who want to linger awhile. The scattered concrete boulders and occasional construction debris make straying from the designated path rather risky.

Carlson Oxbow Park

Carlson Oxbow Park is an eighty-nine-acre tract with trails around a former river bend, now an oxbow lake. It is located off the frontage road that parallels Interstate 80/94 between Kennedy and Cline Avenues. It has a long boardwalk over the wet areas, paved paths through a small wooded area, a few picnic shelters, and fishing piers.[14]

One of Hammond's newer parks, this land was purchased in 1964 by the city of Hammond for flood-control purposes. When the Little Calumet River was straightened in 1975, the former channel became a lake. Instead of draining or developing this largely wet area, the city turned it into a nature park in 1998.

The Carlson Oxbow Park pier on the Little Calumet River.

Miller Woods. *Shirley Heinze Land Trust.*

NATURE PRESERVES IN GARY'S MILLER NEIGHBORHOOD

Miller Woods

Miller Woods, which is part of the National Lakeshore, has guided hikes led by a park ranger every Sunday afternoon during the summer. A fairly recent extension of the Miller Woods trail goes all the way north to the Lake Michigan beach and provides greater access to more of the dune-and-swale habitat, which the public before had seldom seen.

The Paul H. Douglas Center for Environmental Education is located just west of Lake Street. The parking lot, however, is east of the street, and a crosswalk leads to the center itself. Marquette Park, while not a nature preserve, recently underwent a $28 million restoration, which included the restoration of its buildings as well as habitat reconstruction in its natural areas.

Green Heron Pond. *Shirley Heinze Land Trust.*

Green Heron Pond

Green Heron Pond is a small twelve-acre property west of Marquette Elementary School. It contains a high dune and both a permanent pond and seasonal ponds. Black oak savanna and woodland merge with an oak forest in this small preserve. The landscape supports fish, frogs, turtles, muskrats, beavers, and waterfowl. It is owned and maintained by Shirley Heinze Land Trust.

Miller and Bayless Dunes

Miller and Bayless Dunes are part of a sixteen-acre collection of non-contiguous Miller properties along and near Forest Avenue and Indian Boundary Road. Acquired by Shirley Heinze Land Trust via purchase

Miller Dunes. *Shirley Heinze Land Trust.*

and donation, the sites together provide great examples of the wide range of ecosystems—including dry oak forest, interdunal wetland, and blow-outs—found in the high dunes region of Northwest Indiana. Adding to their value are both the rare plants that are found there and the bright colors of the trees every fall.[15]

NATURE PRESERVES IN HOBART

Hobart, one of the oldest cities in Northwest Indiana, still has several large parcels of open space. The most obvious one is Lake George, created in 1845 or '46 when George Earle built a dam across Deep River. Also within the city is the Hobart Marsh–Robinson Lake area east of Interstate 65. Several conservation organizations own and manage na-

Bur Oak Woods. *Shirley Heinze Land Trust.*

ture preserves in this and other areas. Among Hobart's preserves are Bur Oak Woods, Gordon & Faith Greiner Nature Preserve, Cressmoor Prairie Nature Preserve, and the Eleven-Acre Prairie.

Bur Oak Woods

Bur Oak Woods is home to one of Indiana's rarest natural communities, a bur oak savanna. In addition to its woodlands, the site also contains a sedge meadow and some wet depressions. Many of the trees here are several hundred years old, which is quite amazing because after the Chicago and Calumet Area fires of 1871, there was an enormous demand for lumber, resulting in thousands of oak trees being felled in northern Lake County. Yet somehow these wonderful oaks survived the ax and chain-

Cressmoor Prairie.

saw. Funding for the purchase of this eighty-five-acre Shirley Heinze property came from the US Fish and Wildlife Service. There is a trail and a parking lot on Liverpool Road.[16]

Cressmoor Prairie Nature Preserve

Cressmoor Prairie Nature Preserve may be the best-protected black-soil tallgrass prairie in Indiana. The land, though flat and once surrounded by farmland, was never plowed. The preserve has seven threatened species of plants and fifteen threatened species of insects. It is a dedicated Indiana state nature preserve.

The first thirty-six acres were purchased in 1996 with partial funding from the Indiana Heritage Trust. Five more acres have been purchased and received by donation since then. There is a trail and a parking lot on Lake Park Avenue.[17]

Gordon & Faith Greiner Nature Preserve

Gordon & Faith Greiner Nature Preserve is a seventy-three-acre parcel that contains a remnant of a black oak savanna, a stepped fen, peaty wetlands, wet prairie, and a wet woodland habitat. Fields at the south end are being transformed into a tallgrass prairie. The first sixty acres were purchased in 1999 with funds from the US Fish and Wildlife Service. The remaining thirteen acres were purchased the same year at a tax sale.[18]

Eleven-Acre Prairie, Brinson, and Bock/Higgins Nature Preserves

These three parcels, totaling thirty-six acres, are all located within the Hobart Marsh area. Eleven-Acre Prairie is an eleven-acre (of course) remnant mesic prairie that NIPSCO donated to Save the Dunes. The preserve has rare small sundrops, bottled gentian, rose gentian, and green fringed orchids. Brinson and Bock/Higgins are two preserves near Robinson Lake and Hobart Prairie Grove (part of the National Lakeshore) that were previously farm fields. Save the Dunes was able to restore these three properties through the EPA's Great Lakes Restoration Initiative (GLRI). The funding allowed the removal of invasive species and the restoration of native bur oak savanna on the Brinson and Bock/Higgins properties.[19]

NATURE PRESERVES ALONG THE LITTLE CALUMET RIVER

Little Calumet River Prairie and Wetlands Nature Preserve

The Little Calumet River Prairie and Wetlands Nature Preserve is on an eleven-acre strip of land between the Little Calumet River levee and the IU Northwest far north parking lot. The land is owned by the Gary Park Department but it has been restored by IU Northwest personnel, primarily Spencer Cortwright. When restoration work began, 98 percent

The Little Calumet River Preserve north of the IU Northwest campus.

The Little Calumet River Project Area. *Shirley Heinze Land Trust.*

of the site's plant species were Eurasian invasive weeds, with perhaps five to ten native plant species—mainly trees. Today the preserve has about two hundred native plant species.[20]

Little Calumet River Project Area

The Little Calumet River Project Area is composed of two noncontiguous parcels in Porter County's Westchester Township. The properties contain a rich riparian habitat with floodplain and old oxbows. This part of the river is favored by bird enthusiasts, hikers, paddlers, and anglers. A trail that parallels the river can be accessed on the east side of Brummitt Road south of the Little Calumet River Bridge.[21] As of this writing, Shirley Heinze Land Trust is raising funds aimed at connecting protected land parcels along the Little Calumet River, preserving more riparian habitat, and developing additional outdoor recreation opportunities.

NATURE PRESERVES IN PORTAGE AND CHESTERTON

John Merle Coulter Nature Preserve

The John Merle Coulter Nature Preserve is a ninety-two-acre property located at the western edge of the city of Portage on County Line Road. It is adjacent to the National Lakeshore's Tolleston Dunes area. This property was heavily sand mined from the early to the mid-twentieth century. The B&O Railroad extended a rail spur into the area, and by the 1950s the Bos Sand Company was loading and sending out ten railroad cars of sand each day.[22] In spite of all this, the land has reverted to good-quality sand prairie and black oak savanna with more than four hundred species of plants, including nineteen that are endangered. The endangered Karner blue butterfly has previously been seen at this site, but the lupine (which its larvae eat) is threatened by the large number of deer in the area.[23] The old B&O spur has been converted into an east–west hiking trail.

The original sixty-six acres of this preserve were purchased by Shirley Heinze Land Trust in 1997 as an Indiana Heritage Trust project. The Coulter property is now a dedicated state nature preserve. Visitors can

Facing, top: Prickly pear cactus at the Coulter Preserve. *Shirley Heinze Land Trust.*

Facing, bottom: The Coulter Preserve in autumn. *Shirley Heinze Land Trust.*

The Coffee Creek Watershed Preserve. *Shirley Heinze Land Trust.*

enter the preserve from a parking lot east of County Line Road, just south of the MonoSol plant.[24]

Coffee Creek Watershed Preserve

The Coffee Creek Watershed Preserve is a 157-acre parcel containing wetlands, woodlands, and prairies, with more than three miles of trails. The preserve protects more than five hundred species of plants. It is owned and managed by the Coffee Creek Watershed Conservancy, a private group whose mission is to "identify, maintain, and enhance the remnant biodiversity that resides within the corridor while providing educational and recreational experiences for all."[25]

The Great Marsh in Beverly Shores. *Shirley Heinze Land Trust.*

NATURE PRESERVES IN BEVERLY SHORES

Beverly Shores Project Area

The Beverly Shores Project Area is made up of literally hundreds of primarily noncontiguous lots in and south of the town of Beverly Shores, which together total more than eighty acres. Shirley Heinze Land Trust acquired them over the past twenty-five years through donations, tax sales, and other purchases. This project area is part of the Great Marsh, the long, narrow wetland that before settlement paralleled the shoreline of Lake Michigan all the way from Gary to Michigan City. Also located in Beverly Shores are two small rare-plant refuges: McAllister Prairie and Hiawatha Meadow.[26]

NATURE PRESERVES IN MICHIGAN CITY

Michigan City is the oldest, and was for many years the largest, Indiana city on Lake Michigan. The home of Save the Dunes, it incorporates land from Mount Baldy on the west to the Michigan state line. Fortunately, it has a number of preserved parcels. The largest is Ambler Flatwoods, a 353-acre dedicated state nature preserve in Springfield Township.

Barker Woods

Barker Woods is a thirty-acre Shirley Heinze Land Trust dedicated state nature preserve located on Barker Road right near the center of Michigan City. This rare urban old-growth forest community contains pine, oak, dogwood, and maple trees as well as the partridge berry vine and numerous native shrubs. A hiking trail begins in the parking area in front of Barker House (which contains the offices of Save the Dunes). In 1980 the land was bequeathed to The Nature Conservancy by the estate of Margery Barker, and the conservancy transferred ownership to Shirley Heinze Land Trust in 2006.[27]

Stockwell Woods Nature Preserve

This preserve, owned by Save the Dunes, comprises close to thirteen acres of forested foredune and black oak savanna right near Lake Michigan in the town of Long Beach. Save the Dunes has removed invasive species, thinned the tree canopy, and reestablished native herbaceous plants. The preserve, which has state rare dwarf honeysuckle, jack pine, poke milkweed, and box huckleberry, was donated to Save the Dunes by The Nature Conservancy in 2005.[28]

Rakowski Parcel

What had been a small private lot along the Lake Michigan shoreline was donated by Alex and Jude Rakowski of Long Beach to Save the Dunes for long-term preservation. The Rakowski Parcel is special in that it is the only land owned by Save the Dunes that is actually on the Lake Michigan shoreline.

Barker Woods. *Shirley Heinze Land Trust.*

MORE PRESERVATION EFFORTS

Indiana Department of Natural Resources

The Indiana Department of Natural Resources (IDNR) has a number of important properties in addition to Indiana Dunes State Park, including land in the dune-and-swale area of northern Lake County as well as parcels in the Valparaiso Moraine adjacent to the National Lakeshore's Heron Rookery. Its Reynolds Creek Gamebird Habitat Area alone consists of 1,250 acres of land just east of the Porter/LaPorte county line.

Izaak Walton League

The Izaak Walton League of America, Porter County chapter, has obtained more than one hundred acres in several noncontiguous wetland properties southeast of the IDNR properties next to the Heron Rookery. The chapter will restore the lands as a habitat for fish and wildlife. Additional purchases in this area, combined with preserves now owned by the IDNR, Shirley Heinze Land Trust, and the National Park Service, may someday create a contiguous block of land of perhaps 1,900 acres along the upper east branch of the Little Calumet River.[29]

ArcelorMittal

The ArcelorMittal Steel Company worked with the Wildlife Habitat Council (WHC) to conserve and restore more than forty acres of duneland habitat within its Burns Harbor facilities. It then added an employee walking trail. These restoration efforts led to the site's recognition by the WHC as a certified Corporate Lands for Learning site in 2013, and as a Wildlife at Work site the following year.[30]

Notes

Part I. The Unrestricted Use of Natural Resources
1. Cottman, *Indiana Dunes State Park*, 7.
2. "Lakeside Doomed," *Chesterton Tribune*, June 5, 1890.

1. Calumet Beginnings and the Birth of American Ecological Science
1. Chrzastowski and Thompson, "Late Wisconsinan and Holocene Coastal Evolution," 401.
2. Schoon, *Calumet Beginnings*, 38–39.
3. Ibid., 41.
4. Ibid., 28–29.
5. Cook, *Henry Chandler Cowles*, 32.

2. Marquette and the Marquette Plan
1. Danckers et al., *Compendium of the Early History of Chicago*, 252.
2. Moore, *Calumet Region*, 24.
3. Brennan, *Wonders of the Dunes*, 29.
4. Houseal Lavigne Associates, *Marquette Plan*, 1.
5. Northwest Indiana Regional Development Authority, accessed March 12, 2016, http://www.rdatransformation.com.
6. Northwestern Indiana Regional Planning Commission, *Marquette Plan*, 2.
7. Dan Plath, "Lake Michigan Water Trail," Northwest Indiana Paddling Association, accessed March 12, 2016, http://www.nwipa.org/lmwtdedication.html.

3. Natural Resources of the Calumet Area
1. Eenigenburg, *Settlement of the Calumet Region*, n.p.
2. Schoon, *Calumet Beginnings*, 97.
3. Robinson, "History of Lake County, 1833–1847," 60.
4. Packard, *History of LaPorte County*, 33.
5. Eenigenburg, *Settlement of the Calumet Region*, n.p.
6. Unfortunately, Mr. Forsythe soon decided that this operation wasn't necessary and so returned all his books to his home just before fire did strike, burning his books and his home to the ground.

7. Moore, *Calumet Region*, 101.

8. Daniels, *Twentieth Century History*, 168.

9. Bieber and Smith, *Industrial Sands of the Indiana Dunes*, 29.

10. Ball, *Northwestern Indiana from 1800 to 1900*, 350.

11. Schoon, *Calumet Beginnings*, 100.

12. Ibid.

13. Ibid., 101–102.

14. Moore, *Calumet Region*, 194.

15. Howat, *Standard History of Lake County*, 386.

16. Ibid., 460.

17. Berry Lake no longer exists. It was drained by the Standard Oil Company.

18. *Industrial Chicago*, 787.

19. Woods, *First One Hundred Years of Lake County*, 140.

20. "Tolleston Gun Club Now Passing," *Gary Tribune*, May 15, 1908.

21. Schoon, *Calumet Beginnings*, 184.

22. Schoon and Schoon, *Portraits of a Ridge Family*, 20.

23. Cannon, Loring, and Robb, *History of the Lake and Calumet Region*, 178.

24. Woods, *First One Hundred Years of Lake County*, 142.

25. Ibid.

26. Ibid., 142.

27. Ibid., 143.

28. Jackson, *Natural Heritage of Indiana*, 148.

29. Schoon, *Calumet Beginnings*, 46.

30. Brock, *Birds of the Indiana Dunes*, 66–67, 75.

4. Industrialization of the Lakefront

1. David Bensman and Mark R. Wilson, "Iron and Steel," *Encyclopedia of Chicago*, accessed March 13, 2016, http://www.encyclopedia.chicagohistory.org/pages/653.html.

2. Simons and Parker, *Railroads of Indiana*, 33.

3. "Haskell & Barker," Mid-Continent Railway Museum, accessed February 5, 2016, http://www.midcontinent.org/rollingstock/builders/haskellbarker.htm.

4. Simons and Parker, *Railroads of Indiana*, 33.

5. Moore, *Calumet Region*, 197.

6. Ibid., 188.

7. Ibid., 178.

8. "Whiting, Indiana—August 27, 1955."

9. BP, *Whiting Refinery Facility Fact Sheet*.

10. "Whiting, Indiana—August 27, 1955."

11. Moore, *Calumet Region*, 220–224.

12. Morris, *Inland Steel at 100*, 17.

13. LeVan, *East Chicago Diamond Jubilee*, 10.

14. Morris, *Inland Steel at 100*, 45.

15. "Dawes Raises Flag at Opening of Buffington," *Gary Post-Tribune*, June 10, 1927.

16. Morris, *Inland Steel at 100*, 23.

17. Packard, *History of LaPorte County*, 82.

18. Chicago withdrew its suit when U.S. Steel agreed to separate its human sewage and industrial waste and send its human sewage to the Gary sewage treatment plant. The industrial waste continued to be dumped untreated. Hurley, *Environmental Inequalities*, 43.

19. Besozzi, "Hammond Water Filtration Plant," n.p.

20. Ibid.

5. Industrialization of the Grand Calumet River and the Indiana Harbor Ship Canal

1. Moore, *Calumet Region*, 146–147.

2. Ibid., 150–151.

3. Trusty, *Hammond*, 23.

4. Moore, *Calumet Region*, 152.

5. Trusty, *Hammond*, 26.

6. Moore, *Calumet Region*, 160.

7. Ibid., 240–241.

8. "Grand Calumet River Area of Concern," Indiana Department of Environmental Management, accessed March 13, 2016, http://in.gov/idem/cleanwater/2424.htm.

9. Hurley, *Environmental Inequalities*, 34.

10. Bukro, "Indiana Harbor Plants Turning Lake into Sewer for Industry," *Chicago Tribune*, September 17, 1967.

11. "Grand Calumet River/Indiana Harbor Canal," US Department of the Interior, accessed March 30, 2016, http://www.cerc.usgs.gov/orda_docs/CaseDetails?ID=71.

6. The Push for Parks and Duneland Development, 1890–1929

1. Maloney, *Prairie Club of Chicago*, 8.

2. Ibid.

3. Chew, "100 Years of the Prairie Club."

4. Schoon, *Dreams of Duneland*, 208.

5. Ibid., 209.

6. Chew, "100 Years of the Prairie Club."

7. Schoon, *Dreams of Duneland*, 208.

8. Chew, "100 Years of the Prairie Club."

9. Maloney, *Prairie Club of Chicago*, 121.

10. Schoon, *Dreams of Duneland*, 212.

11. Ibid.

12. Meister, Martin, and the Historical Society of Ogden Dunes, *Ogden Dunes*, 33.

13. Schoon, *Dreams of Duneland*, 228–229.

14. Ibid., 167.

15. Moore, *Calumet Region*, 12.

16. Schoon, *Dreams of Duneland*, 220.

7. Port versus Park

This chapter has been condensed from "Port or Park: Dreams in Dissonance," in Schoon, *Dreams of Duneland*, 235–240.

1. Engel, *Sacred Sands*, 255.

2. Read, "Saving the Dunes," 2.

Part II. Returning to Sustainability

1. Indiana Department of Environmental Management, *Remedial Action Plan*, introduction.

2. Ibid.

3. The Superfund program is a federal government effort to clean up land that has been contaminated by hazardous waste that poses a risk to human health and/or to the environment.

4. United Nations World Commission on Environment and Development, *Our Common Future*, 16.

5. "IDEM's Mission Statement," Indiana Department of Environmental Management, accessed March 14, 2016, http://www.in.gov/idem/5215.htm.

8. Earth Consciousness in the '60s and '70s

1. US Environmental Protection Agency, Office of Water, *Report to Congress*, 2–1.

2. Transcript of audiotapes of the Apollo 8 telecast, December 24, 1968, p. 274, US National Archives, Washington, DC.

3. Time-Life Books, *100 Photographs That Changed the World*, 172.

4. "Earthrise, Christmas Eve 1968," Dill Knob Observatory, accessed March 14, 2016, http://dillknobobservatory.net/wnc-skies/2008–2/earthrise-christmas-eve-1968 (search for Earthrise 1968).

5. Larry Phillips, "Remembering Scoop Jackson," *Seattle Times*, September 2, 2008, http://www.seattletimes.com/opinion/remembering-scoop-jackson.

6. Wertz, *Lynton Keith Caldwell*, 132.

7. Liroff, "NEPA Litigation in the 1970s," 316.

8. Hurley, *Environmental Inequalities*, 136–137.

9. "40th Anniversary of the Clean Air Act," US Environmental Protection Agency, accessed March 31, 2016, https://www.epa.gov/clean-air-act-overview/40th-anniversary-clean-air-act.

10. "Evolution of the Clean Air Act," US Environmental Protection Agency, accessed March 14, 2016, http://www.epa.gov/clean-air-act-overview/evolution-clean-air-act.

11. Ibid.

12. Richard M. Nixon, "Reorganization Plan No. 3 of 1970," special message to Congress, July 9, 1970, https://www.epa.gov/aboutepa/reorganization-plan-no-3–1970.

13. "EPA History: Agency Accomplishments," US Environmental Protection Agency, accessed March 14, 2016, http://www.epa.gov/aboutepa/epa-history-agency-accomplishments.

14. "EPA: A Retrospective, 1970–1990," US Environmental Protection Agency, November 29, 1990, http://www2.epa.gov/aboutepa/epa-retrospective.

15. "The History of Earth Day," Earth Day Network, accessed December 25, 2015, http://www.earthday.org/earth-day-history-movement.

16. Bill Christofferson, "The First Earth Day: April 22, 1970," *The Xoff Files* (blog), April 22, 2006, http://thexofffiles.blogspot.com/2006/04/first-earth-day-april-22–1970.html.

17. Hurley, *Environmental Inequalities*, 19.

18. "History of the Clean Water Act," US Environmental Protection Agency, accessed March 14, 2016, https://www.epa.gov/laws-regulations/history-clean-water-act.

19. "Canada-US Great Lakes Water Quality Agreement Overview," Environment and Climate Change Canada, last modified July 5, 2013, https://www.ec.gc.ca/grands lacs-greatlakes/default.asp?lang=En&n=E615A766–1.

20. Ibid.

21. "Restoring the Great Lakes Areas of Concern," Environment and Climate Change Canada, last modified February 22, 2016, https://www.ec.gc.ca/indicateurs -indicators/default.asp?lang=en&n=E08EA691-1.

9. The Road to Cleaner Air

1. Easterly, *Northwest Indiana Partners for Clean Air,* slide 10.

2. Ibid., slide 12.

3. "Progress Cleaning the Air and Improving People's Health," US Environmental Protection Agency, accessed January 12, 2016, http://www.epa.gov/clean-air-act -overview/progress-cleaning-air-and-improving-peoples-health.

4. "Six Common Pollutants: Lead in Air; Health," US Environmental Protection Agency, accessed March 15, 2016, http://www3.epa.gov/airquality/lead/health.html.

5. "Six Common Pollutants: Nitrogen Dioxide," US Environmental Protection Agency, accessed March 15, 2016, http://www3.epa.gov/airquality/nitrogenoxides.

6. "Particulate Matter (PM-10)," US Environmental Protection Agency, last modi- fied February 22, 2016, http://www3.epa.gov/airtrends/aqtrnd95/pm10.html.

7. BP, *Whiting Business Unit,* 3.

8. Ibid.

9. Morris, *Inland Steel at 100,* 59.

10. "Air, Land and Water," ArcelorMittal, accessed June 15, 2015, http://usa .arcelormittal.com/Corporate-responsibility/Environment/Air-land-water.

11. Ibid.

12. NIPSCO, *2014 Integrated Resource Plan,* 10.

13. "Improving Air Quality," NIPSCO, accessed December 17, 2015, https://www .nipsco.com/about-us/our-environment/improving-air-quality.

14. Ibid.

15. Maureen Groppe, "Indiana Power Plants Set for Pollution Limits," *Indianapolis Star,* June 1, 2014, http://www.indystar.com/story/news/politics/2014/06/01/indiana -power-plants-ready-pollution-limits-thanks-rate-hikes/9787315.

16. NIPSCO, *2014 Integrated Resource Plan,* 10.

17. "Environmental Stewardship," U.S. Steel, accessed August 29, 2015, https://www .ussteel.com/uss/portal/home/aboutus/environment/company-environmental +stewardship.

18. U.S. Steel, *Gary Works,* 16.

10. The Road to Cleaner Water

1. Indiana Department of Environmental Management, *Indiana Integrated Water Monitoring and Assessment Report,* 10.

2. Greer, "Obstacles to Taming Corporate Polluters," 200.

3. Jon C. Schladweiler, "The History of Sanitary Sewers: Part 9; Disposal of Sani- tary Sewage," accessed March 15, 2016, http://www.sewerhistory.org/time-lines /tracking-down-the-roots-of-our-sanitary-sewers/part-9-disposal-of-sanitary-sewage.

4. Stevenson Swanson, "Reversing the River," *Chicago Tribune*, July 18, 2008, http:// www.chicagotribune.com/news/nationworld/politics/chi-chicagodays-reversingriver -story-story.html.

5. Schladweiler, "History of Sanitary Sewers."

6. Robert B. Semple Jr., "Happy Birthday, Clean Water Act," *Taking Note* (blog), *New York Times*, October 16, 2012, http://takingnote.blogs.nytimes.com/2012/10/16 /happy-birthday-clean-water-act.

7. US Environmental Protection Agency, Office of Water, *Report to Congress*, ES-4.

8. Don Woodard, e-mail to the author, November 2015.

9. In the sixteen months before the basins were put into service, the same three stations discharged raw sewage into the Grand Calumet River a total of 162 times.

10. US Environmental Protection Agency and Indiana Department of Environmental Management, *Northwest Indiana Environmental Initiative Action Plan*, 1.

11. Federal Water Pollution Control Administration, *Proceedings in the Matter of Pollution*, 15.

12. Morris, *Inland Steel at 100*, 59.

13. City of Gary, *Gary Comprehensive Plan*, 2.

14. Dorreen Carey, conversation with the author, June 30, 2015.

15. Joyce Russell, "Making a Clean Sweep," *Times*, October 2, 2004.

16. Impairments include total fish consumption restrictions, beach closings, fish tumors or deformities, reproductive problems, and loss or degradation of fish and wildlife habitats.

17. "Habitat Restoration," Illinois-Indiana Sea Grant Program, accessed March 29, 2016, http://www.greatlakesmud.org/habitat-restoration.html.

18. "Great Lakes Legacy Act," Illinois-Indiana Sea Grant Program, accessed March 23, 2016, http://www.greatlakesmud.org/great-lakes-legacy-act.html.

19. Ibid.

20. Officially the Great Lakes–St. Lawrence River Basin Water Resources Compact.

21. "Great Lakes Water Resources Compact and Agreement," Alliance for the Great Lakes, accessed March 14, 2016, http://www.greatlakes.org/Page.aspx?pid=1330 &frcrld=1.

22. "Water Quality in Indiana: Citizens Advisory for the Remediation of the Environment (CARE) Committee," Indiana Department of Environmental Management, accessed March 16, 2016, http://in.gov/idem/cleanwater/2425.htm.

23. "Water Quality in Indiana: Grand Calumet River Area of Concern," Indiana Department of Environmental Management, accessed March 16, 2016, http://in.gov/idem /cleanwater/2424.htm.

24. Christine Kraly, "Rivers Run through Dredging," *Times*, April 20, 2008, http:// www.nwitimes.com/special-section/news/rivers-run-through-dredging/article _75a10f74-f98b-5135-b261–8427031a9cdc.html.

25. Kari Lydersen, "A Toxic River Improves, but Still Has Far to Go," *New York Times*, June 2, 2011, http://www.nytimes.com/2011/06/03/us/03cncriver.html.

26. Riffles are rocky shoals or sandbars below the surface of a waterway. They usually produce ripples. Because of the resultant higher turbulence, they oxygenate the stream water. The ripples also make it more difficult to see into or out of the water and so they form a shelter from predators.

27. McMurray et al., *Biological Community*, 36.

28. "Legacy Act Cleanup of Grand Calumet River," US Environmental Protection Agency, last modified August 24, 2015, https://www.epa.gov/grand-calumet-river-aoc /legacy-act-cleanup-grand-calumet-river.

29. Ingersoll et al., "Toxicity Assessment of Sediments," 156.

30. Illinois-Indiana Sea Grant, "Nature Is Right Down the Street for East Chicago Students," *Lakeside Views* (blog), May 18, 2015, http://lakesideviews.blogspot.com/2015/05/nature-is-right-down-street-for-east.html.

31. "Water Quality in Indiana: Grand Calumet River Area of Concern."

32. Lydersen, "Toxic River Improves."

33. Ibid.

34. "General Information," Hammond Sanitary District, accessed March 16, 2016, http://www.gohammond.com/departments/sanitary-district/general-information.

35. McMurray et al., *Biological Community*, 36.

36. "Clean Water Act Celebrates 40 Years of Water Improvements," MWH Global, October 1, 2012, http://www.mwhglobal.com/thought-leadership/water-management /clean-water-act-celebrates-40-years-water-improvements.

37. James Salzman, "Why Rivers No Longer Burn," *Slate*, December 10, 2012, http://www.slate.com/articles/health_and_science/science/2012/12/clean_water_act_40th _anniversary_the_greatest_success_in_environmental_law.html.

38. Semple Jr., "Happy Birthday, Clean Water Act."

11. Lake Michigan Health, Beach Closures, and Fishing

1. "About Our Great Lakes: Great Lakes Basin Facts," National Oceanic and Atmospheric Administration, Great Lakes Environmental Research Laboratory, accessed March 16, 2016, http://www.glerl.noaa.gov/pr/ourlakes/facts.html.

2. Hurley, *Environmental Inequalities*, 42.

3. Angela Howe and Mara Dias, "The BEACH Act at 15 Years," Surfrider Foundation, March 31, 2015, http://www.surfrider.org/coastal-blog/entry/the-beach-act-at-15 -years.

4. BeachGuard, Indiana Department of Environmental Management, accessed March 29, 2016, https://extranet.idem.in.gov/beachguard.

5. "Testing the Waters: A Guide to Water Quality at Vacation Beaches; State Summary; Indiana," Natural Resources Defense Council, accessed December 25, 2015, http://www.nrdc.org/water/oceans/ttw/in.asp.

6. Ibid.

7. State of the Beach/State Reports/IN/Water Quality," Beachapedia, last modified June 22, 2015, http://www.beachapedia.org/State_of_the_Beach/State_Reports/IN /Water_Quality.

8. Schoon, *Calumet Beginnings*, 153–160.

9. Packard, *History of LaPorte County*, 95.

10. Daniels, *Twentieth Century History*, 164–165.

11. "Sea Lamprey Control in the Great Lakes," Great Lakes Fishery Commission, accessed March 16, 2016, http://www.glfc.org/sealamp/how.php.

12. Ibid.

13. Charlebois, "Nonindigenous Threats Continue," 5.

14. Howard Meyerson, "The Salmon Experiment: The Invention of a Lake Michigan Sport Fishery, and What Has Happened Since," *Grand Rapids Press*, April 18, 2011, http://www.mlive.com/outdoors/index.ssf/2011/04/the_salmon_experiment_the_inve.html.

15. "Creek Information," Trail Creek Guide Service, accessed March 16, 2016, http://www.trailcreekguideservice.com/creek.htm.

16. Ibid.

17. Brian Breidert, e-mail to the author, June 3, 2015.

18. "Trail Creek Lamprey Barrier to Be Dedicated," Indiana Department of Natural Resources, April 18, 2012, http://www.in.gov/activecalendar_dnr/EventList.aspx?view=EventDetails&eventidn=5674&information_id=11394&type=&syndicate=syndicate.

19. "Fish Granted Protection Near New Sea Lamprey Barrier," Indiana Department of Natural Resources, October 26, 2012, http://www.in.gov/activecalendar_dnr/EventList.aspx?view=EventDetails&eventidn=5264&information_id=10527&type=&syndicate=syndicate.

12. Brownfields Restored to Usefulness

1. "Brownfield Overview and Definition," US Environmental Protection Agency, accessed March 17, 2016, https://www.epa.gov/brownfields/brownfield-overview-and-definition.

2. "ROI Results 2014," 2.

3. "Success in Hammond: Former Myers and West Point Industrial Park Properties Redeveloped," Indiana Finance Authority, August 2005, http://www.in.gov/ifa/brownfields/files/Myers_West_Point.pdf.

4. *Munster Steel*.

5. Trusty, *Munster*, 44.

6. Ronald Robbins and Jeanne Robbins, conversations with the author, May 2015.

7. Chelsea Schneider Kirk, "Munster Steel Breaks Ground on New Facility in Hammond," *Times*, October 9, 2013, http://www.nwitimes.com/news/local/lake/hammond/munster-steel-breaks-ground-on-new-facility-in-hammond/article_f42e4ad5-fb7e-5b17-9320-4a17a8ee21f7.html.

8. "The Lost Marsh Environmental and Recreation Area," Northwest Indiana Brownfields Coalition, accessed March 19, 2016, http://www.nwibrownfields.com/projects.

9. Ibid.

10. Ibid.

11. Ibid.

12. Schoon, *Calumet Beginnings*, 177.

13. Trusty, *Munster*, 77.

14. Ibid., 137.

15. Ibid., 79.

16. Ibid., 208.

17. LuAnn Franklin, "Centennial Park Becomes Munster's Crown Jewel of Parks," *Times*, June 16, 2009.

18. Energy Systems Group, *Centennial Park*, 8.

19. Garard, "Hobart Township," 527.

20. Northwest Indiana Regional Planning Commission, *Implementation*, 75.

21. "Portage Lakefront and Riverwalk," US Department of the Interior, accessed March 19, 2016, https://www.doi.gov/greening/awards/2011/idnl.

22. Ibid.

23. Ibid.

24. "Portage Lakefront Pavilion," Design Organization, accessed March 19, 2016, http://www.designorg.com/project/portage-lakefront-pavilion.

25. "Portage Lakefront and Riverwalk."

26. John Robbins, "Big Plans to Expand, Restore Lakefront Park in Portage," *Indiana Economic Digest*, April 14, 2014, http://indianaeconomicdigest.com/main.asp?Section ID=31&SubSectionID=79&ArticleID=74278.

27. Carole Carlson, "Portage Lakefront Park Restoration Continues," *Gary Post-Tribune*, December 16, 2015.

28. "Superfund 35th Anniversary," US Environmental Protection Agency, last modified February 1, 2016, http://www.epa.gov/superfund/superfund-35th-anniversary.

29. Lauri Harvey Keagle, "Health Concerns at Center of EC Lead, Arsenic Cleanup," *Times*, September 4, 2014, http://www.nwitimes.com/news/local/lake/east-chicago /health-concerns-at-center-of-ec-lead-arsenic-cleanup/article_bbf6467a-e346–56fd -ac13–7a8948e0aa33.html.

13. Solid Waste and Recycling

1. Trusty, *Munster*, 109–110.

2. Ibid.

3. "Keyes Fibre Company History," Funding Universe, accessed March 19, 2016, http://www.fundinguniverse.com/company-histories/keyes-fibre-company-history.

4. Kathy Kassabaum, "Recycling: Should Communities Begin?," *Times*, September 25–26, 1991.

5. Watson, "Curbside Goes High-Tech."

6. Kassabaum, "Recycling."

7. Trusty, *Munster*, 168.

8. Tracy Hack, "Many Area Towns Taking Spin on Recycling Bandwagon," *Times*, August 7, 1991.

9. Tim Summers, "Recycling in Lake County," *Times*, November 4, 1991.

10. Kassabaum, "Recycling."

11. Summers, "Recycling in Lake County."

12. Indiana Legislative Services Agency, *Issues Relating to Recycling*, i.

13. Ibid., 103.

14. Recycling and Waste Reduction District of Porter County, *Strategic Plan*, 11.

15. "ReUZ Room," Lake County Solid Waste Management District, accessed March 19, 2016, http://www.lcswmd.com/reuzroom.htm.

16. "Paint Recycling," City of Hobart, accessed March 19, 2016, http://www .cityofhobart.org/index.aspx?nid=196.

17. "Curbside Recycling," LaPorte County Solid Waste District, accessed March 29, 2016, http://solidwastedistrict.aradisehost.com/programs/curbside.html.

18. "Recycling Center," Homewood Disposal, accessed March 19, 2016, http://mydisposal.com/general/recycling-center.

19. Jeanette Romano, e-mail to the author, July 2015.

14. Local Pioneering Environmental Heroes and Heroines

1. Cassidy, "Henry Chandler Cowles," 12.

2. Cook, *Henry Chandler Cowles*, 24.

3. Ibid., 67.

4. Cockrell, *Signature of Time and Eternity*, 4–5.

5. Schoon, *Dreams of Duneland*, 188.

6. Cook, *Henry Chandler Cowles*, 60–63.

7. Dabbert, *Indiana Dunes Revealed*, 3.

8. Ibid., 12.

9. Ibid., 17.

10. Ibid., 39.

11. The Campbell Settlement House, at 2244 Washington Street in Gary, was a project of the National Women's Missionary Society to the Methodist Church. It ran a day nursery, and held classes for adults that stressed interracial harmony and spiritual and educational training.

12. Cottman, *Indiana Dunes State Park*, 37.

13. Cockrell, *Signature of Time and Eternity*, 4–5.

14. Engel, *Sacred Sands*, 256.

15. Ibid., 255.

16. Quoted in Engel, *Sacred Sands*, 254.

17. Engel, *Sacred Sands*, 255.

18. Ibid., 258.

19. "Dorothy Richardson Buell," Find a Grave, June 28, 2012, http://www.findagrave.com/cgi-bin/fg.cgi?page=gr&GRid=92745769.

20. Douglas, *In the Fullness of Time*, 13.

21. Engel, *Sacred Sands*, 229–230.

22. Ibid., 6.

23. Wertz, *Lynton Keith Caldwell*, 28.

24. Helen Walcher was also the daughter of a manager at the American Steel Foundries plant in Hammond.

25. Bartlett and Gladden, "Lynton K. Caldwell and Environmental Policy."

26. "Prophet with Honor Lays Down Pen," *Times*, June 7, 1990, http://www.nwitimes.com/uncategorized/prophet-with-honor-lays-down-pen/article_7163805a-90a8-585d-bbe7-4f0472d9d724.html.

15. Environmental Education Opportunities

1. Cowles, *Plant Societies of Chicago and Vicinity*, 75.

2. Ibid.

3. Northwest Indiana Restoration Monitoring Inventory, accessed March 19, 2016, http://nirmi.org.

4. Argyilan, *Building Educational Collaborations*.

5. "Illinois-Indiana Sea Grant Extension Office," Purdue University Calumet, accessed March 20, 2016, http://webs.purduecal.edu/ems/sea-grant.

6. Dunes Learning Center, accessed March 20, 2016, http://duneslearningcenter.org.

7. Schoon, *Dreams of Duneland*, 256. Adapted from an essay by Lee Botts.

8. "About Us," Shirley Heinze Land Trust, accessed March 20, 2016, http://www.heinzetrust.org/about-us.html.

9. Ibid.

10. "Mighty Acorns," Shirley Heinze Land Trust, accessed March 20, 2016, http://www.heinzetrust.org/mighty-acorns.html.

11. Ibid.

12. Amber Horbovetz, Dunes Learning Center lead naturalist, correspondence with the author, July 2015.

13. "Sandhill Cranes Fall Migration," Indiana Department of Natural Resources, accessed March 20, 2016, http://www.in.gov/dnr/fishwild/3109.htm.

14. Brad Bumgardner, e-mail to the author, July 2015.

15. Lake County Solid Waste Management District, accessed March 20, 2016, http://www.lcswmd.com/ourservices.htm.

16. Recycling and Waste Reduction District of Porter County, accessed March 20, 2016, http://www.itmeanstheworld.org.

17. Solid Waste District of LaPorte County, accessed March 20, 2016, http://www.solidwastedistrict.com. The Visual Recycling Guide can be found at http://solidwastedistrict.aradisehost.com/downloads/22316-SolidWaste-flyer.pdf.

18. "Merit Badges," Boy Scouts of America, accessed March 20, 2016, http://www.scouting.org/scoutsource/BoyScouts/AdvancementandAwards/MeritBadges.aspx.

19. "Insignia List," Girl Scouts of the United States of America, accessed March 20, 2016, http://www.girlscouts.org/en/our-program/uniforms/insignia-list.html.

16. Preservation and Restoration of Natural Areas

1. Burnham and Bennett, *Plan of Chicago*, 50.

2. Schoon, *Calumet Beginnings*, 215.

3. Schoon, *Dreams of Duneland*, 223.

4. Ibid., 227.

5. Noel B. Pavlovic, e-mail to the author, September 9, 2015.

6. Reshkin, "Natural Resources of the Calumet," 5.

7. "Indiana Wetland Program Plan," Indiana Department of Environmental Management, accessed March 20, 2016, http://www.in.gov/idem/wetlands/2334.htm.

8. "Cowles Bog Restoration Project," National Park Service, accessed March 20, 2016, http://www.nps.gov/indu/learn/nature/great-marsh-restoration.htm.

9. National Park Service, *Great Marsh Restoration at Indiana Dunes National Lakeshore*.

10. "Welcome to Gibson Woods Nature Preserve and Environmental Awareness Center," Lake County Parks, accessed March 20, 2016, http://www.lakecountyparks.com/gibson.html.

11. "Great Lakes Legacy Act," Illinois-Indiana Sea Grant Program, accessed March 23, 2016, http://www.greatlakesmud.org/great-lakes-legacy-act.html.

12. Paul Labus, conversation with the author, June 2015.

13. Semko, "Hammond Lakefront Park & Bird Sanctuary."

14. Bob Sweet, "Family Trips—Oxbow Park," Examiner.com, June 20, 2010, http://www.examiner.com/article/family-trips-oxbow-park.

15. Shirley Heinze Land Trust, *Guidebook to the Nature Preserves*, 40.

16. Ibid., 16.

17. Ibid., 20.

18. Ibid., 30.

19. "Managing Our Lands," Save the Dunes, accessed March 20, 2016, http://savedunes.org/managing-our-lands.

20. "Little Calumet River Prairie and Wetlands Preserve," Indiana University Northwest, accessed March 20, 2016, http://www.iun.edu/coas/related-information/little-calumet-river-prairie-and-wetlands-preserve.htm.

21. Shirley Heinze Land Trust, *Guidebook to the Nature Preserves*, 62–63.

22. Schoon, *Sand Mining*, 57.

23. Schoon, *Dreams of Duneland*, 270.

24. Shirley Heinze Land Trust, *Guidebook to the Nature Preserves*, 52–53.

25. Coffee Creek Watershed Preserve, accessed March 20, 2016, http://www.coffeecreekwc.org.

26. Shirley Heinze Land Trust, *Guidebook to the Nature Preserves*, 48.

27. Ibid., 78–80.

28. "Managing Our Lands."

29. Jim Sweeney, e-mail to the author, August 2015.

30. "Air, Land and Water," ArcelorMittal, accessed June 15, 2015, http://usa.arcelormittal.com/Corporate-responsibility/Environment/Air-land-water.

Bibliography

Argyilan, Erin. *Building Educational Collaborations in Northwest Indiana through GLIS-TEN*. PowerPoint presentation. Harrisburg, PA: GLISTEN, 2010. https://www.csu .edu/cerc/documents/ArgyilanErin-GLISTEN.pdf.

Ball, Timothy Horton. *Lake County, Indiana, from 1834 to 1872*. Chicago: J. W. Good-speed, 1873.

———. *Northwestern Indiana from 1800 to 1900*. Chicago: Donohue and Henneberry, 1900.

Bartlett, Robert V., and James N. Gladden. "Lynton K. Caldwell and Environmental Policy: What Have We Learned?" In *Environment as a Focus for Public Policy*, by Lynton K. Caldwell, 3–23. Edited by Robert V. Bartlett and James N. Gladden. College Station: Texas A&M University Press, 1995.

Bennett, Ira Elbert. *History of the Panama Canal: Its Construction and Builders*. Washington, DC: Historical Publishing, 1915.

Besozzi, Leo. "Hammond Water Filtration Plant." In *Hammond Indiana Water Filtration Plant*. Hammond, IN: Department of Water Works, 1936.

Bieber, C. L., and Ned M. Smith. *Industrial Sands of the Indiana Dunes*. Bulletin No. 7. Bloomington: Indiana Geological Survey, 1952.

Blatchley, Willis S. *Geology of Lake and Porter Counties, Indiana*. 22nd Annual Report of the Department of Geology and Natural Resources of Indiana. Indianapolis: Department of Geology and Natural Resources of Indiana, 1897.

BP. *Whiting Business Unit: Environmental Statement for Year 2012 (Review of Y2011 Performance)*. Whiting, IN: BP, 2012.

———. *Whiting Refinery Facility Fact Sheet*. Whiting, IN: BP, 2015.

Brennan, George A. *The Wonders of the Dunes*. Indianapolis: Bobbs-Merrill, 1923.

Brock, Kenneth J. *Birds of the Indiana Dunes*. Rev. ed. Michigan City, IN: Shirley Heinze Environmental Fund, 1997.

Brown, Steven E., N. K. Bleuer, and Todd A. Thompson. *Geologic Terrain Map of the Southern Lake Michigan Rim, Indiana*. Bloomington: Indiana Geological Survey, 1996.

Burnham, Daniel H., and Edward H. Bennett. *Plan of Chicago*. New York: Princeton Architectural Press, 1993. First published 1908 by Commercial Club of Chicago. Page references are to the 1993 edition.

Caldwell, Lynton K. "Environment: A New Focus for Public Policy?" *Public Administration Review* 23 (1963): 132–139.

Calvert, Gwalter C. "Hoosier Slide." Unpublished manuscript, undated. Michigan City (Indiana) Historical Society.

Cannon, Thomas H., H. H. Loring, and Charles J. Robb. *History of the Lake and Calumet Region of Indiana.* Indianapolis: Historians' Association, 1927.

Cassidy, Victor M. "Henry Chandler Cowles: Ecologist, Teacher, Conservationist." *Chicago Wilderness*, Spring 2007.

Charlebois, Patrice. "Nonindigenous Threats Continue." *The Helm* 13, no. 1 (1996): 5–7.

Chew, Ryan. "100 Years of the Prairie Club." *Chicago Wilderness*, Spring 2008.

Chrzastowski, Michael J., and Thompson, Todd A. "Late Wisconsinan and Holocene Coastal Evolution of the Southern Shore of Lake Michigan." *Quaternary Coasts of the United States: Marine and Lacustrine Systems*, SEPM Special Publication no. 48 (1992): 397–413.

Cockrell, Ron. *A Signature of Time and Eternity: The Administrative History of Indiana Dunes National Lakeshore.* Omaha, NE: National Park Service, 1988.

Commonwealth Biomonitoring. *Pathogen Assessment: Grand Calumet / Little Calumet Rivers, Hammond, Indiana; A Report to the Hammond Sanitary District.* Indianapolis: Commonwealth Biomonitoring, 2002.

Cook, Sarah Gibbard. *Henry Chandler Cowles (1869–1939) and Cowles Bog, Indiana: A Study in Historical Geography and the History of Ecology.* Chicago: Field Museum of Natural History, 1980. Reprint, Field Museum of Natural History, 1999.

Cottman, George Streiby. *Indiana Dunes State Park: A History and Description.* Indianapolis: Indiana Department of Conservation, 1930.

Cowles, Henry C. *The Plant Societies of Chicago and Vicinity.* Chicago: University of Chicago Press, 1901.

Dabbert, James R., ed. *The Indiana Dunes Revealed: The Art of Frank V. Dudley.* Champaign: University of Illinois Press, 2006.

Danckers, Ulrich, Jane Meredith, John F. Swenson, and Helen Hornbeck Tanner. *A Compendium of the Early History of Chicago: To the Year 1835 When the Indians Left.* River Forest, IL: Early Chicago, 2000.

Daniels, E. D. *A Twentieth Century History and Biographical Record of LaPorte County, Indiana.* Chicago: Lewis, 1904.

Douglas, Paul H. *In the Fullness of Time: The Memoirs of Paul H. Douglas.* New York: Houghton Mifflin Harcourt, 1972.

Easterly, Thomas W. *Northwest Indiana Partners for Clean Air.* PowerPoint presentation. Indianapolis: Indiana Department of Environmental Management, 2013. http://www.in.gov/idem/files/commish_pres_20130418_pca.ppt.

Eccles, William John. *The French in North America, 1500–1783.* Markham, ON: Fitzhenry & Whiteside, 1998.

Eccleston, Charles H. *NEPA and Environmental Planning: Tools, Techniques, and Approaches for Practitioners.* Boca Raton, FL: CRC Press, 2008.

———, ed. *The NEPA Planning Process: A Comprehensive Guide with Emphasis on Efficiency.* New York: Wiley, 1999.

Eenigenburg, Harry. *The Settlement of the Calumet Region.* Lansing, IL: privately printed, ca. 1941.

Energy Systems Group. *Centennial Park, Munster, IN.* Promotional packet. Chicago: Energy Systems Group, 2015.

Engel, J. Ronald. *Sacred Sands: The Struggle for Community in the Indiana Dunes.* Middletown, CT: Wesleyan University Press, 1983.

Federal Water Pollution Control Administration. *Proceedings in the Matter of Pollution of the Interstate Waters of the Grand Calumet River: Indiana and Illinois.* Washington, DC: US Department of the Interior, 1969.

Federal Water Pollution Control Administration, Office of Public Information, Great Lakes Region. *Clean Water for Mid-America.* Chicago: Federal Water Pollution Control Administration, 1970.

Garard, George A. "Hobart Township." In *The Counties of Porter and Lake,* edited by Weston A. Goodspeed and Charles Blanchard, 522–532. Chicago: F. A. Battery, 1882.

Gary, Indiana, City of. *Gary Comprehensive Plan—2008–2028.* Gary, IN: City of Gary, 2008.

Goodspeed, Weston A., and Charles Blanchard, eds. *The Counties of Porter and Lake.* Chicago: F. A. Battery, 1882.

Gray, Ralph D. *Public Ports for Indiana: A History of the Indiana Port Commission.* Indianapolis: Indiana Historical Bureau, 1998.

Greer, Edward. "Obstacles to Taming Corporate Polluters: Water Pollution Politics in Gary, Indiana." *Boston College Environmental Affairs Law Review* 3, no. 2 (1974): 199–220.

Houck, Oliver A. "More Net Loss of Wetlands: The Army-EPA Memorandum of Agreement on Mitigation Under the §404 Program." *Environmental Law Reporter,* June 1990. http://elr.info/sites/default/files/articles/20.10212.htm.

Houseal Lavigne Associates. *The Marquette Plan.* Prepared for the Northwestern Indiana Regional Planning Commission and the Indiana Department of Natural Resources. Chicago: Houseal Lavigne Associates, 2008.

Howat, William Frederick. *A Standard History of Lake County, Indiana, and the Calumet Region.* Chicago: Lewis, 1915.

Hurley, Andrew. *Environmental Inequalities: Class, Race, and Industrial Pollution in Gary, Indiana, 1945–1980.* Chapel Hill: University of North Carolina Press, 1995.

Indiana Department of Environmental Management. *Indiana Integrated Water Monitoring and Assessment Report to the U.S. EPA.* Indianapolis: Indiana Department of Environmental Management, 2014. http://www.in.gov/idem/nps/files/ir_2014_report.pdf.

———. *The Remedial Action Plan for the Indiana Harbor Canal, the Grand Calumet River and the Nearshore Lake Michigan: Stage One.* Indianapolis: Indiana Department of Environmental Management, 1991.

———. *25th Anniversary: State of the Environment 2011.* Indianapolis: Indiana Department of Environmental Management, 2011. http://www.in.gov/idem/files/state_of_environment_2011.pdf.

Indiana Department of Natural Resources, Lake Michigan Fisheries Office. *History of Lake Michigan Fisheries.* Accessed December 25, 2015. http://www.in.gov/dnr/fishwild/files/fw-History_Lake_Michigan_Fisheries.pdf.

Indiana Legislative Services Agency, Recycling Evaluation Committee. *Issues Relating to Recycling and Solid Waste Management Programs.* Indianapolis: Indiana Legislative

Services Agency, 2003. http://www.in.gov/legislative/pdf/recyclingsolidwaste
    webdoc.pdf.

Industrial Chicago: The Building Interests. Chicago: Goodspeed, 1891.

Ingersoll, C. G., D. D. MacDonald, W. G. Brumbaugh, B. T. Johnson, N. E. Kemble, J. L.
    Kunz, T. W. May, et al. "Toxicity Assessment of Sediments from the Grand Calumet
    River and Indiana Harbor Canal in Northwestern Indiana, USA." *Archives of Environ-
    mental Contamination and Toxicology* 43, no. 2 (2002): 156–167.

Jackson, Marion T. *The Natural Heritage of Indiana.* Bloomington: Indiana University
    Press, 1997.

Jones, W. D. *Northern Portions of Lake County, Indiana.* Chicago: Rufus Blanchard, 1906.

*Lake County Blue Book, 1897–98: A Review of Lake County, Indiana, as to County and City
    Officials, Etc., for the Current Year.* Chicago: Umbdenstock, 1898.

Lake County Solid Waste Management District. *Solid Waste Management Plan.* Crown
    Point, IN: Lake County Solid Waste Management District, 1993.

LeVan, Rose G. *East Chicago Diamond Jubilee Historical Record.* East Chicago, IN: pri-
    vately printed, 1968.

Liroff, Richard A. "NEPA Litigation in the 1970s: A Deluge or a Dribble?" *Natural Re-
    sources Journal* 21 (April 1981): 315–330.

Maloney, Cathy Jean. *The Prairie Club of Chicago.* Chicago: Arcadia, 2001.

Manny, Carter Hugh. "Hoosier Slide." Unpublished manuscript of an undated speech,
    ca. 1960. Old Lighthouse Museum, Michigan City, Indiana.

McMurray, Paul D., Jr., James R. Stahl, Anne Kominowski, and James R. Smith. *Biologi-
    cal Community, Contaminants and Toxicity Monitoring on the Grand Calumet River and
    Indiana Harbor Ship Canal Area of Concern, 2013.* Indianapolis: Indiana Department
    of Environmental Management, 2013.

McShane, Stephen G., and Gary S. Wilk. *Steel Giants: Historic Images from the Calumet
    Regional Archives.* Bloomington: Indiana University Press, 2009.

Meister, Dick, Ken Martin, and the Historical Society of Ogden Dunes. *Ogden Dunes.*
    Chicago: Arcadia, 2001.

Metcalf, Leonard, and Harrison P. Eddy. *American Sewerage Practice.* Vol. 1. New York:
    McGraw-Hill, 1914.

Mighty Acorns. Curriculum Guide: Levels One–Three. Chicago: Mighty Acorns, 2015.

Moore, Powell A. *The Calumet Region: Indiana's Last Frontier.* Indianapolis: Indiana His-
    torical Bureau, 1959.

Morris, Jack H. *Inland Steel at 100: Beginning a Second Century of Progress.* Des Plaines,
    IL: Bradley Printing, 1993.

Mumford, Russell E. "Wings across the Sky: Birds of Indiana." In *The Natural Heritage
    of Indiana*, edited by Marion T. Jackson, 329–339. Bloomington: Indiana University
    Press, 1997.

Munster Steel. Promotional flyer. Munster, IN: Munster Steel, 2014.

National Oceanic and Atmospheric Administration and Indiana Department of Natural
    Resources. *Combined Coastal Program Document and Final Environmental Impact
    Statement for the State of Indiana.* Silver Spring, MD: National Oceanic and Atmo-
    spheric Administration; Indianapolis: Indiana Department of Natural Resources,
    2002. http://www.in.gov/dnr/lakemich/files/lmcp-feis.pdf.

National Park Service. *Great Marsh Restoration at Indiana Dunes National Lakeshore.* Undated flyer.

NIPSCO. *2014 Integrated Resource Plan Executive Summary.* Merrillville, IN: NIPSCO, 2014. https://www.nipsco.com/docs/default-source/about-nipsco-docs/2014-nipsco-irp-executive-summary.pdf.

Northwestern Indiana Regional Planning Commission. *Implementation.* Portage, IN: Northwestern Indiana Regional Planning Commission, 2007. http://nirpc.org/media/57250/Implementation.pdf.

————. *The Marquette Plan: Regional Projects.* Portage, IN: Northwestern Indiana Regional Planning Commission, 2015. http://www.nirpc.org/media/55113/mp_2015_regional_projects_final.pdf.

Oglesbee, Rollo B., and Albert Hale. *History of Michigan City, Indiana.* Evansville, IN: Edward J. Widdell, 1908.

One Region. *Northwest Indiana Profile: 2012 Quality of Life Indicators Report.* Munster, IN: One Region, 2012.

Packard, Jasper. *History of LaPorte County, Indiana, and Its Townships, Towns, and Cities.* LaPorte, IN: S. E. Taylor, 1876.

Pavlovic, Noel B., and Marlin L. Bowles. "Rare Plant Monitoring at Indiana Dunes National Lakeshore." In *Science and Ecosystem Management in the National Parks,* edited by William L. Halvorson and Gary E. Davis, 253–280. Tucson: University of Arizona Press, 1996.

Perry, W. A. "A History of Inland Steel Company." Unpublished manuscript, ca. 1978. Calumet Regional Archives.

Read, Herbert. "Saving the Dunes: A Battle Waged over Decades." Undated typewritten manuscript. Calumet Regional Archives, Chesterton, Indiana.

Recycling and Waste Reduction District of Porter County. *Strategic Plan.* Valparaiso, IN: Recycling and Waste Reduction District of Porter County, 2015.

Reshkin, Mark. "The Natural Resources of the Calumet: A Region Apart." In *Sand and Steel: The Dilemma of Cohabitation in the Calumet Region,* edited by Stephen G. McShane, 3–6. Gary: Indiana University Northwest, 1987.

Robinson, Solon. "History of Lake County, 1833–1847." In *History of Lake County,* vol. 10, by the Lake County Historical Association, 35–67. Gary, IN: Calumet Press, 1929.

"ROI Results 2014." *Indiana Brownfields Bulletin,* Winter 2015. http://www.in.gov/ifa/brownfields/files/FINAL_1–28–2015%282%29.pdf.

Schoon, Kenneth J. *Calumet Beginnings: Ancient Shorelines and Settlements at the South End of Lake Michigan.* Bloomington: Indiana University Press, 2003.

————. *Dreams of Duneland: A Pictorial History of the Indiana Dunes Region.* Bloomington: Indiana University Press, 2013.

————. "Natural Richness and Industrial Power." *Chicago Wilderness,* Spring 2009.

————. *Sand Mining in and around the Indiana Dunes National Lakeshore.* Washington, DC: National Park Service, 2015.

Schoon, Kenneth J., and Margaret S. Schoon. *Portraits of a Ridge Family: The Jacob Schoons.* Munster, IN: privately printed, 1981.

Semko, Laura. "Hammond Lakefront Park & Bird Sanctuary." *Chicago Wilderness,* Fall 2007.

Shirley Heinze Land Trust. *Guidebook to the Nature Preserves of Shirley Heinze Land Trust*. Valparaiso, IN: Shirley Heinze Land Trust, 2015.

Simmons, J. W., C. R. Lee, D. L. Brandon, H. E. Tatem, and J. G. Skogerboe. *Information Summary, Area of Concern: Grand Calumet River, Indiana*. Miscellaneous Paper EL-91–10. Vicksburg, MS: US Army Engineers Waterways Experiment Station, 1991.

Simons, Richard S., and Francis H. Parker. *Railroads of Indiana*. Bloomington: Indiana University Press, 1997.

Sinko, Peggy Tuck. "Hammond: A Centennial Portrait by Lance Trusty." *Indiana Magazine of History*, March 1987.

Solow, Robert. "An Almost Practical Step toward Sustainability." In *Assigning Economic Value to Natural Resources*, by the National Research Council, 19–29. Washington, DC: National Academy Press, 1994.

Time-Life Books. *100 Photographs That Changed the World*. New York: Life Books, 2003.

Trusty, Lance. *Hammond: A Centennial Portrait*. Norfolk, VA: Donning, 1984.

———. *Munster: A Centennial History*. Norfolk, VA: Donning, 2006.

United Nations World Commission on Environment and Development. *Our Common Future*. New York: United Nations, 1987. http://www.un-documents.net/our-common-future.pdf.

US Army Corps of Engineers and US Environmental Protection Agency. *Indiana Harbor and Canal Maintenance Dredging and Disposal Activities Comprehensive Management Plan*. Vol. 2, *Technical Appendices*. Chicago: US Army Corps of Engineers and US Environmental Protection Agency, 1998.

US Environmental Protection Agency, Office of Air and Radiation. "Air Quality Trends." Washington, DC: US Environmental Protection Agency, Office of Air and Radiation. Accessed July 1, 2016. https://www3.epa.gov/airtrends/aqtrends.html #airquality.

US Environmental Protection Agency and Indiana Department of Environmental Management. *Northwest Indiana Environmental Initiative Action Plan*. Chicago: US Environmental Protection Agency and Indiana Department of Environmental Management, 1996.

US Environmental Protection Agency, Office of Water. *Report to Congress: Combined Sewer Overflows to the Lake Michigan Basin*. Washington, DC: US Environmental Protection Agency, Office of Water, 2007. https://www3.epa.gov/npdes/pubs/cso _reporttocongress_lakemichigan.pdf.

U.S. Steel. *Gary Works*. Promotional brochure. Gary, IN: undated.

Watson, Tom. "Curbside Goes High-Tech." *Resource Recycling*, October 1990. http://infohouse.p2ric.org/ref/03/02932.pdf.

Wertz, Wendy Read. *Lynton Keith Caldwell: An Environmental Visionary and the National Environmental Policy Act*. Bloomington: Indiana University Press, 2014.

"Whiting, Indiana—August 27, 1955." *Industrial Fire World* 16, no. 2 (2001). http://fireworld.com/Archives/tabid/93/articleType/ArticleView/articleId/87066/Whiting-Indiana—August-27-1955.aspx.

Woods, Sam B. *The First One Hundred Years of Lake County, Indiana*. Crown Point, IN, 1938.

Writers' Program of the Work Projects Administration. *The Calumet Region Historical Guide*. Gary, IN: Garman, 1939.

Index

Page numbers in *italics* refer to figures and tables.

KENNETH J. SCHOON is Professor Emeritus of Science Education at Indiana University Northwest and a Northwest Indiana native. In 1990, after twenty-two years as a middle and high school science teacher, he joined the IU Northwest faculty, retiring twenty-three years later. His research interests have centered on science misconceptions and local studies. In addition to his publications on misconceptions and urban education, he is author of *Calumet Beginnings: Ancient Shorelines and Settlements at the South End of Lake Michigan* (IUP, 2003); *Dreams of Duneland: A Pictorial History of the Indiana Dunes Region* (IUP, 2013); and *City Trees: ID Guide to Urban & Suburban Species* (2007). He is a founding board member and past president of the Dunes Learning Center, a member of the Advisory Council of Shirley Heinze Land Trust, a board member of the Munster Education Foundation, and a member of the IU Northwest Science Olympiad steering committee.

This book was designed by Pam Rude and set in type by Tony Brewer at Indiana University Press and printed by Four Colour Imports Ltd.

The fonts are Arkhip, designed by Igor Kuznetsov in 2014; Arno, designed by Robert Slimbach in 2007; Candida, designed by Jakob Erbar in 1935; and Dancing Script, designed by Pablo Impallari in 2011. Arkhip and Dancing Script are available from their designers. Arno and Candida were published by Adobe Systems Incorporated.